ACCOUNTABILITY IN URBAN SOCIETY

Public Agencies Under Fire

Volume 15, URBAN AFFAIRS ANNUAL REVIEWS

INTERNATIONAL EDITORIAL ADVISORY BOARD

ACCOUNTABILITY IN URBAN SOCIETY

Public Agencies Under Fire

Edited by

Scott Greer

Ronald D. Hedlund

and

James L. Gibson

Volume 15, URBAN AFFAIRS ANNUAL REVIEWS

SAGE PUBLICATIONS Beverly Hills London

For information address:

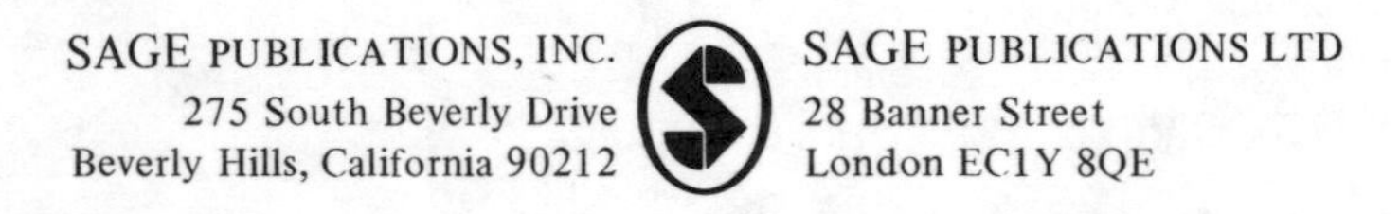

SAGE PUBLICATIONS, INC.
275 South Beverly Drive
Beverly Hills, California 90212

SAGE PUBLICATIONS LTD
28 Banner Street
London EC1Y 8QE

Printed in the United States of America

Library of Congress Cataloging in Publication Data

Main entry under title:

Accountability in urban society.

(Urban affairs annual reviews ; v. 15)
Includes bibliographical references.
1. Municipal government–United States–Addresses, essays, lectures. 2. Municipal services–United States–Addresses, essays, lectures. 3. Political participation––United States–Addresses, essays, lectures.
I. Greer, Scott A. II. Hedlund, Ronald D., 1941-
III. Gibson, James L. IV. Series.
HT108.U7 vol. 15 [JS341] 301.36'08s [320]
ISBN 0-8039-1081-9 78-17295
ISBN 0-8039-1082-7 pbk.

FIRST PRINTING

CONTENTS

PREFACE

☐ A MAJOR REASON for having a university-based research center is to provide a source of uncontaminated information and a neutral arena in which major issues may be addressed. For this reason the Urban Research Center of the University of Wisconsin, Milwaukee, sponsored a Conference on Institutional Accountability in Urban Society in the spring of 1977. Although this is an issue of great saliency today, it is not one which elicits broad academic concern; thus we felt it was a part of the Center's mission to focus attention on the problems of accountability.

The Conference stimulated a considerable intellectual interest which has resulted in a number of essays on the subject, and in this volume. I am sure all the contributors are aware of the complexities of the problem and of the difficulties in conceptualizing it; this volume can only be a beginning, an opening of a serious discussion. On one hand, custom has lagged far behind our emerging dilemmas; on the other hand, intentionally crafted social devices for guaranteeing accountability are as likely as not to formalize the techniques of evasion.

At the same time we consider accountability we must be concerned with the protection of properly placed trust; it is the lubrication that allows the society to act. Indeed, trust is the means for transforming accountability into individual responsibility, without which no governance can be effective and equitable. Thus our problem is far more than that of setting up an infinite regress of auditing agencies; it is one of creating a moral culture for a high technology, large-scale society which is, in the nature of things, delegated.

These essays address only a part of the problem; they are heavily biased toward the public realm. This is because most writers who address the problem have the government in mind. It is, however, worth remembering that a very large part of our actions which should be held accountable are in the private realm, from commerce to the family, and that these actions count heavily. Corporate thievery and inept parenting are both enemies of the society; no policemen can control them, but an awareness of the public's concern through social devices may keep them within bounds. Torn between *laissez faire* and the social contract, we continue the task of the American citizen—to do the impossible.

May I extend my personal gratitude to George Keulks, Dean of the Graduate School at the University of Wisconsin, whose interest in this topic was a source of early inspiration and continued support. Thanks are due also to the State Legislative Leader's Foundation and to the Wisconsin Center for Public Policy, groups which helped to make the conference possible, and most importantly to the editors of this book who brought the amorphous topic with which we began into focus with this volume.

—Ann Lennarson Greer
Urban Research Center
The Graduate School
University of Wisconsin-Milwaukee

Introduction:
The Accountability of Institutions in Urban Society

SCOTT GREER
RONALD D. HEDLUND
JAMES L. GIBSON

□ ONE CAN FIND THE ROOTS OF CONCERN with accountability in the predictions of Max Weber and others long ago. Though Weber typically overestimated the efficiency of the formally organized society, "the bureaucratic solution," he nevertheless pointed out the main danger; that in the interests of the existing bureaucratic structure and its goals, the very meanings and values of the larger society would be jettisoned. Certainly there are many indications today that a large and influential body of notables and common citizens are disturbed by the imputed lack of accountability–read this irresponsibility–of the very public organs and publicly certified professions which are supposed to care for their needs.

In the practice of medicine today we note the rapid increase in legal action, in findings for the layman, and in the size of awards for malpractice. This, together with the strong governmental move to establish mandatory peer-review among medical practitioners and the strong thrust for responsible health planning by hospitals and others at the local level, indicate widespread dissatisfaction with the way these ancient licensed and subsidized public servants are behaving. The same is true for the law, where "legal malfeasance and malpractice" are becoming well known concepts. And, in lesser degree, even the practitioners of higher education are under fire and in the courts over such matters as truth in labeling, unfair discrimination for ethnic reasons, and malfeasance. (Malpractice has not yet been alleged, since nobody knows just how important formal education is to the development of competence.)

All of these developing trends reflect a discontent with the terms of trade. They indicate a strong belief among many that they are the victims of broken bargains—by doctors, hospitals, attorneys, and professors.

The same trend occurs in the movement toward a more rigorous standard of performance by the legislative, judicial, and executive branches of government at every level. Nader's Raiders are only one edge of a movement which criticizes the practice of governance: from the demand for "sunshine laws" in legislatures to insistence on "zero base budgeting" in bureaucratic agencies, an influential section of the citizenry is concerned with performance and accountability. Again, one suspects that they are unhappy with the terms of trade, disturbed by *broken bargains.*

One can think of a number of reasons for this growing distrust of duly constituted authority. Certainly the breakdown of a community, in which the subcommunities of government and the professions once not only had integrity, but had an interleaving integration with the broader middle-class society, is critical. Much malpractice adjudication can be explained in this way. The sheer scale of the society and of the public agencies is certainly a contributory factor: the citizen becomes anonymous client and source of revenue to the agencies and they become vested interests to him. At the same time, the increasing internal inclusiveness of a mass democracy leads to a wide distribution of "rights" among populations which, in the past, were held politically inert by exclusion on grounds of class, race, sex, or age. The agencies and professions lose their middle-class base of support, while new challenges come from the once excluded, the "insulted and injured" of this society.

Thus there is a widespread demand for a better bargain, better terms of trade between agency and citizen, and *one which can be enforced.* The question is: How to legislate and administer accountability? It is a bureaucratic solution to problems posed by formal organization of basic functions; we move from a set of morals which are no longer relevant or effective toward a set of rules which we hope will cure the malaise brought about by rule-ridden groups and organizations. We use bureaucratic strategies on the bureaucrats.

In doing so, however, we find that every strategy has serious flaws. We try to evaluate the quality of service by structure, by process, and by output. In the structural approach we ask: are the necessary resources present in sufficient numbers? The buildings, machinery, chemicals, and bodies with appropriate certificates? And if they are—still the question arises: what difference does it make? In the processual evaluation we ask: did the various actors do all they were supposed to, without wasting resources on the superfluous? Did they give the test, check the statute, handle the legal cases on time, consider and act on the bills before them? And if they did—still the question: what difference? The third strategy of accountability holds the actor responsible for the *outcome* of his actions. This is the most critical matter of all: only from gauging outcomes can we measure what the agency produced. Only as it relates *to* outcome does either structure or process standard have validity. Yet to evaluate outcome is the most difficult type of evaluation. We must know exactly what we expect, and control for a host of variables which affect the difficulty of producing it.

In our society, with its disintegrated class system and folk society, we adopt bureaucratic strategies to cope with the problems of controlling formal organization. In that case we had better keep in mind the immense capacity of social beings to cope with new controls, to manipulate them, trick them, and turn them to their own ends. One can imagine new structures of accountability which would allow rote performance, an expensive acquisition of structural furniture, an expensive performance of unnecessary procedures, for the simple purpose of protecting "doctor, lawyer, merchant, thief."

At the same time, the residual commitment to duty and concern with outcome, whether legislative program or medical case, may be further eroded by the intrusion of such rules. Thus, ironically, the introduction of new modes of accountability may increase needless activity and expense and destroy trust, while evading the basic problem they were meant to solve. Once more, social mechanisms may produce results 180 degrees away from what was intended.

The problem, then, is to prevent the self-fulfilling prophecy which, assuming accountability is lost, uses rules of accounting which effectively lock the door through which the virtue disappeared. How shall we approach this problem?

First, we believe that for any system of maintaining or increasing accountability to have the effects intended, it must be more than a maze constructed to channel individual behavior. At key points in certain roles there must be a personal, moral engagement with the rules of the game, and these must be at least cognate with the morality of most people involved as clients, citizens, or governors. In short, between the principle and mechanism of accountability and its end product, justice, there must be responsibility in the individual and structured group. Furthermore, between formal accountability and individual responsibility, there is another key link: trust. Any effective system of accountability must regard social trust as the basic resource for social responsibility on all sides.

The exciting thing about the move toward greater accountability is that it recalls very basic moral demands of human society. The frightening aspect of the moment is the possibility that it may further destroy our stores of social trust which, unlike the economist's postulated "free givens," are socially earned and can be socially destroyed in short order.

In attempting to deal with these demands for accountability, the public sector first needs to consider several perplexing questions:

- What is accountability?
- To whom are public officials and agencies accountable?
- For what actions should the public sector be held accountable?
- Under what conditions should the private sector be subjected to accountability controls?
- What are the problems inherent in understanding accountability?
- How can progress be made in understanding accountability through research?
- What is a reasonable research agenda for studying accountability?

The chapters in this book do not address all of these questions; nor are the solutions proposed definitive. Nevertheless, as an initial step in that direction, they attempt to foster dialogue and provide tenable directions for the public sector in responding to this crucial concern.

Part I

Accountability and Professionalism in the Delivery of Services

Introduction

□ A LINCHPIN IN MANY ARGUMENTS progressing from the way things are done to the accountability of the doer is the notion of professionalism. This term has been applied to occupations as far apart as physicians, Chicago Flat Janitors, and athletes. It includes notions ranging from "playing for payment" to that of a self-regulating collegium which ensures that all members will place the code of practice above all else. Eliot Friedson argues that "the most strategic distinction lies in legitimate organized autonomy–that a profession is distinct from other occupations in that it has been given the right to control its own work" (Friedson and Lorber, 1972: 71). In return, the public has a right to assume that the members of the profession will establish and maintain the highest standards of work.

In this section the professions are examined from the viewpoint of those concerned with accountability to the public. Michael Lipsky, in his discussion of "street level bureaucrats," deals with "low level" workers in the human services who, despite their status, must exercise high degrees of discretion. Such workers cannot be held accountable by bureaucratic control methods, he argues; therefore it is much more to the point to encourage their vocation, their professional commitment to doing their work well. Efforts at bureaucratic control, he argues, are not only inappropriate but are apt to reduce the quality of the services and increase the real cost of the enterprise. The complex struggle among civic departments, public service unions, and struggling administrators does not produce a clear picture of accountability, but it does give a clearer notion of the problems faced in achieving it.

Scott Greer, in his discussion of professional self-regulation in the practice of medicine, operates in a more clinical fashion. First he details the reasons for the present dissatisfaction with the cost, quality, and accessibility of medical services. The technological revolution in medicine, the enormous shift in societal norms which changed medical care from a privilege to a right, and the over-all transformation of urban society are some of his concerns. Looking at the past grounds for confidence in the medical system, he concludes that they cannot be easily reintroduced. Instead, he examines at some length the radical governmental innovation, the Professional Standards Review Organizations. Using general propositions from social science, he attempts to estimate the degree to which the remedy is appropriate for the malady. Can a program mandated by the government but administered by those it is meant to regulate actually produce publicly acceptable accounting? As cynics might say, can we solve the ageless problem of who shall guard the guardians by simply making some of the guards delegates of governmental power? His solution rests upon an improvement in social theory and innovations in social practice.

REFERENCE

FRIEDSON, E., and LORBER, J. [eds.] (1972). *Medical men and their work.* Chicago: Aldine, Atherton.

1

The Assault on Human Services: Street-Level Bureaucrats, Accountability, and the Fiscal Crisis

MICHAEL LIPSKY

□ THIS CHAPTER EXAMINES the current application of administrative measures to secure accountability among lower level workers in certain public agencies. I argue that bureaucratic accountability is virtually impossible to achieve among lower level workers who exercise high degrees of discretion, at least where qualitative aspects of the work are involved. Nonetheless, public managers are pressured to secure or improve workers' accountability through manipulation of incentives and other aspects of job structure immediately available to them. When considered along with other objectives public managers seek, the result is not simply ineffectiveness but an erosion of the foundations of service quality.

People are accountable when there is a high probability that they will be responsive to legitimate authority or influence. This definition of accountability directs attention to two important aspects of the concept. First, accountability is a relationship between people or groups. One is always accountable *to* someone (or groups), never in the abstract. Although the term is sometimes used loosely confusion results unless we specify both parties in the accountability relationship.[1]

Second, accountability refers to patterns of behavior. Only if a pattern of behavior exists can predictability, and therefore accountability, exist. In practical terms this means that efforts to change or improve accountability cannot succeed unless patterns of behavior change or improve. For example, medical review boards and civilian police review boards will not increase accountability unless general relationships with patients and citizens change. This is no more than saying that laws are only effective if they not only punish transgressions but also deter illegal behavior generally.

From this perspective, attempts to increase accountability through administrative controls may be seen as efforts to increase the congruence between worker behavior and the preferences of agency executives through the use of sanctions and incentives available to the organization. However, to utilize organizational incentives and sanctions, at least the following conditions must prevail. These conditions are the prerequisites of a bureaucratic accountability *policy*.

1. Agencies must know what they want workers to do. Where the objectives are multiple, they must be able to rank their preferences.
2. Agencies must know how to measure workers' performance.
3. Agencies must be able to compare workers to one another, to establish a standard for judgment.
4. Agencies must have incentives and sanctions capable of disciplining workers. They must be able to prevail over other incentives and sanctions that may operate.

Manipulation of administrative controls is not the only way to secure accountability. Recent efforts and speculation have also focused on improving accountability through recreating the conditions of a market economy (for example, voucher proposals), changing the governance of programs (for example, school decentralization), and through seeking judicial relief. Considerable emphasis has also been placed recently on improving accountability by enhancing employees' professional training, status, and (thus) obligations to service provision.

Of all these, efforts to obtain bureaucratic accountability are most important. They represent the range of actions thought to be available to people who manage public agencies, and represent the normal route to governmental accountability. Reformers may have other ideas, but public managers normally have only the tools of bureaucratic accountability to apply, as they wait for the dust of reformers to settle.

The preconditions of an accountability policy may exist in many bureaucratic contexts, but there is an area of public policy in which they do not. These are the contexts in which public policy consists of interactions between public employees and citizens, and in which public employees have broad discretion in taking action or making decisions. Bureaucratic accountability policies in these contexts tend to undermine rather than enhance service quality. Efforts to improve accountability may systematically *decrease* service quality when certain conditions of public bureaucracy prevail.

STREET-LEVEL BUREAUCRATS

When tasks are routine, easily specified and measured, public employees' deviation from supervisors' expectations may be treated as questions of organizational "slack." In such cases it might be appropriate to focus on manipulating organizational incentives and sanctions in order to close the gap between expectations and performance. However, accountability poses vastly different problems, and requires an entirely different frame of reference, when low-level employees have discretion in their work, and when the subjects of their work (the citizens who are the subjects of the interactions), are themselves reactive to what happens to them in the interactions.

I have called such public employees "street-level bureaucrats" to draw attention to the fact that this is a generic role in public service delivery. It is occupied by teachers, police officers, welfare workers, public lawyers and judges, some health workers, and other public employees whose interactions with citizens constitute the policy to be delivered.

Street-level bureaucrats may be defined as public employees who exercise considerable discretion in their work and interact with citizens in the course of their jobs about matters of significance to citizens. Their work situations exhibit similar characteristics. They chronically lack the resources to do their jobs as they are generally articulated, in part because the demand for their efforts, attention, time, and services never eases. There is always another case to process, another client to interview. If service resources increase, pressures develop to serve more clients or reduce the backlog of unmet client needs. (In some situations the resource constraints are inherent in the work context. For example, policemen, emergency room personnel, and others with episodic but intense involvement with clients can only marginally increase the information necessary to make a proper decision about taking action.)[2]

Any workers faced with insufficient resources to do their jobs will redefine the jobs or cope with them by developing routines and other adaptive work patterns.[3] Resource constraints are characteristic of many jobs and hardly separate street-level bureaucrats from other workers. However, supervisors in street-level bureaucracies are less able to dictate the nature of the coping mechanisms developed in the work situation, because of other characteristics of street-level work. These include the following:

(1) Street-level bureaucrats have a high degree of discretion in their interactions with citizens (mentioned above).

(2) Agency goals in street-level bureaucracies tend to be ambiguous, vague, or contradictory.

(3) Performance measures are extremely difficult if not impossible to construct when the quality of human interaction is the thing to be measured. And in situations in which performance measures would be possible to obtain, the costs of doing so are prohibitive.

(4) Street-level bureaucrats' clients are not voluntary; clients interact with these agencies unwillingly or because they have no choice if they want to receive significant public goods. Finally, clients are not a primary reference group for street-level bureaucrats. Their expectations, in other words, do not primarily define street-level roles.

Each of these conditions of work is significant because each represents a potential focus of obtaining accountability closed off by the structure of street-level work. If goals were more focused, administrators would be able to orient workers' practice in more uniform directions. If performance measures were available, administrators could more easily sanction unsatisfactory performance. If clients were voluntary, their withdrawal from the active client population might provide a potent signal that the agency as a whole was not performing adequately, inducing workers and administrators alike to provide more satisfactory service to retain clients.[4] If clients were among workers' primary reference groups, their expectations would by definition be incorporated into workers' roles. These conditions of work help explain the relative autonomy of street-level bureaucrats from supervisory or client control. Not only do street-level bureaucrats enjoy considerable discretion in their work, but their autonomy from close guidance is enhanced because some of the common preconditions for obtaining workers' compliance typically are lacking in these agencies.

Notwithstanding (and largely because of) their relative freedom from supervision, considerable attention has recently been devoted to making street-level bureaucrats more accountable. Such efforts draw energy from two related general sources. First, street-level bureaucrats provide significant public goods (through their services), and citizens are affected by and react to the interactions they experience. Thus public attention tends to focus upon the adequacy of performance and the implicit decisions in service delivery. The stakes and potential for abuse of discretionary authority are high. For example, teachers affect the expectations of children and thus their performance; their classifications help determine educational futures. Police discretionary responses to citizens may result in severe injury.[5]

Second, street-level bureaucrats consume a major portion of the budgets of state and local governments. In a period of budgetary restraint and fiscal conservatism, interest in issues of accountability are raised in the context of concerns over the effectiveness of public service work and the return to the public welfare of public service expenditures.[6]

To be sure, these concerns are a proper focus of governmental attention. Indeed, issues of bureaucratic accountability are central concerns of any inquiry into the functioning of modern democracies. However, there are some troubling dimensions to current bureaucratic efforts to increase or insure accountability in street-level bureaucracies. They are troubling because accountability in street-level bureaucracies cannot be achieved through manipulation of bureaucratic incentives; yet efforts to achieve accountability through bureaucratic intervention continue, with destructive implications for the quality of public services.

This is not to say that bureaucratic controls never significantly insure accountability among street-level bureaucrats. Obviously the behavior of all public employees can be oriented in desired directions by administrators. But the closer one gets to issues of service quality, the less appropriate are the standard bureaucratic influences. The issue is not simply that the efforts to obtain accountability are ineffective; they often undermine service delivery, in part because they are spuriously related to quality control. Consider the difficulties for improving bureaucratic accountability posed by the following observations of street-level bureaucrats' work context.

ACCOUNTABILITY TO CLIENTS

The essence of street-level bureaucracies is that they require people to make decisions about other people. Street-level bureaucrats have discretion because the nature of service-provision calls for human judgment which cannot be programmed and for which machines cannot substitute. Street-level bureaucrats have responsibility for making unique and fully appropriate responses to individual clients and their situations. It is the nature of what we call human services that the unique aspects of people and their situations will be apprehended by public service workers and translated into courses of action responsive to each case, within (more or less broad) limits imposed by their agencies. They will not, in fact, dispose of every case in unique fashion. The limitations on possible responses are often circumscribed, for example, by the prevailing statutory provisions of the law, or the categories of welfare services to which recipients can be assigned. However, street-level bureaucrats still have the responsibility *at least to be open to the possibility* that each client presents special circumstances and opportunities which may require fresh thinking and flexible action.

If this is the case, street-level bureaucrats must irreducibly be accountable to the client and for an appropriate response to the client's situation and circumstances. These cannot sensibly be translated into authoritative agency guidelines. It is a contradiction in terms to say that the worker should be accountable to an agency to respond to each client in unique fashion appropriate to the presenting case. For no accountability can exist if the agency does not know what response it prefers, and it cannot assert a preferred response if each worker should be open to the possibility that unique and fresh responses are appropriate. It is more useful to suggest that street-level bureaucrats are ordinarily expected to be accountable to two sources of influence—agency preferences *and* clients' claims.[7]

There are other sources for the assertion that street-level bureaucrats are ordinarily expected to be accountable to clients, in possible opposition to the agencies for which they work. The most important of these is that most street-level bureaucrats are professionals or work in occupations aspiring to professional status. In either case a fundamental expectation attached to the job is that client needs are primary and that the extension of public trust depends upon reciprocal accountability to people as individuals, when they are encountered in the course of work. Social workers, teachers, and of course doctors and lawyers, are expected to respond to the individual and the presenting situation, however much their work situations mitigate against flexible responses.

This is a great strength and also a great weakness of the public services. It provides a measure of responsiveness to clients when the organization of bureaucratic service tends toward neglect or rigidity. But, by virtue of providing another focus of accountability, it also means that street-level bureaucrats are less controllable.

HOLDING WORKERS TO AGENCY OBJECTIVES

Despite the dual focus of accountability inherent in the street-level bureaucrats' roles, public managers are drawn to making street-level bureaucrats more accountable by reducing their discretion and constraining their alternatives. They write manuals to cover contingencies. They audit the performance of workers to provide retrospective sanctions in anticipation of which it is hoped future behavior will be modified. They insist workers specify objectives in the hopes that accountability can be more effectively monitored. These management tools at times may be effective in controlling workers. Manuals specifying proper procedures may help standardize responses and provide instruction. Performance audits may create greater awareness that management is observing performance, and may thus lead to workers taking greater care. Specifying objectives is always likely to be instructive, and to direct workers' attention to the relationship between the available resources and the goals they are trying to achieve.[8]

However, street-level bureaucrats may subvert efforts to control them more effectively in the name of accountability. In these and other examples of attempts to increase control it is relatively easy for workers to tailor their behavior to avoid accountability. For one thing, they are likely to be the source of information management receives concerning their performance. They are fully able to provide information about the presenting situation which makes the action taken appear to be responsive to the original problem when it may not be. This involves blatant falsification less than auspicious shading of the truth and sincere rationalization.

It is extremely difficult for management to contradict workers' reports, for several reasons. A critical piece of information is the state of mind of the worker and his or her analysis of the presenting situation. Since street-level decisions are made in private, it is extremely difficult to second-guess workers, since the second guessers are not at hand to evaluate the intangible factors which may have contributed to the original judgment. The records kept by street-level bureaucrats are almost never complete or adequate to the task of post-hoc auditing for this reason, and when records are kept, they are written sketchily and defensively to guard against later adverse scrutiny.[9]

Record-keeping can help insure that certain procedures are followed (since falsification is normally not the issue). Health practitioners can be made to run certain tests, social workers to ask certain questions, police officers to follow certain procedures. But the records cannot force accountability on the appropriateness of the actions to the presenting situation.

Another major difficulty with obtaining accountability through management control efforts arises because of the dependence of street-level bureaucracies on their workers. Since the services delivered by schools, police departments, or legal services offices fundamentally consist of the actions of teachers, police officers, and lawyers, these agencies are constrained from controlling workers too much, particularly in challenging their performance, for fear of generating opposition to management policies and diminishing accountability even further. The weakness of management incentives to sanction negative performance contributes to a climate in which vigorous challenges to street-level bureaucrats' autonomy in decision-making is presumed to have possible negative net consequences for service delivery, by destroying morale and inhibiting worker initiative.[10]

Are there negative aspects to management control efforts, or are these efforts simply generally ineffective? There are several respects in which such practices can actively subvert service quality.

First, specification of methods of client treatment, under the guise of obtaining accountability, may actually result in reductions in client services. There is often a thin line between inducing workers' behavior to better conform to agency preferences, and inducing workers to be open to fewer options and opportunities for clients. For example, attempts by the Nixon Administration to increase welfare employees' accountability by auditing their error rate in accepting clients for welfare reduced services by providing incentives for welfare workers only to reduce errors which favored clients. Federal guidelines did not call for reduction in error rates for the potential welfare population as a whole. If it had, the applications of all who applied for welfare would have been audited, weighing equally those accepted and those rejected. Scrutiny of welfare workers' decisions strictly in terms of whether or not they were too lenient amounts to narrowing the role of welfare workers, reducing their accountability to clients and to professional standards of conduct. The important point is not that welfare rolls were tightened, but that tightening was accomplished in the name of accountability.

Second, supervision of subordinates with broad discretion and responsibilities requires assertions of priorities in attempting to increase accountability. Police departments may scrutinize traffic tickets, vice arrests, or interracial encounters between police and citizens. But they cannot meaningfully hold officers accountable for everything all the time. If everything is scrutinized, nothing is. Thus efforts to control street-level bureaucrats not only affect those areas which are management targets, but also those areas which are not the focus of management efforts, since by implication those efforts will not come up for surveillance. The danger is that efforts to increase accountability in some areas may come to be regarded as the only areas in which accountability will be sought and behavior scrutinized.

Third, many management control efforts provide a veneer of accountability without in fact constraining behavior very much. Management control systems have symbolic value, providing concerned publics with reassurances that employees are accountable even when they are not. Introduction of management systems at least temporarily permits deflection of criticism of street-level bureaucrats' behavior as citizens find it very difficult to challenge the emperor whom officials say is fully clothed, appearances and personal experiences to the contrary notwithstanding.[11]

GOAL CLARIFICATION

One of the conspicuous features of many public services is the ambiguity and multiplicity of objectives. How can accountability be achieved, ask the critics, if public officials are unclear about their objectives? The desirability of clarifying (and then operationalizing) agency objectives in order to increase accountability stems from the force of this observation, and the recognition that a bureaucratic accountability policy requires specification of objectives (as suggested above).

It is difficult to take issue with the desirability of clarifying objectives if they are needlessly and irrelevantly fuzzy or contradictory. Surely it is easier to run an effective agency if you know what you are supposed to be doing. However, the management orientation to clarifying objectives raises an important issue for public service quality.

The issue is this: agency goals may be unclear or contradictory for reasons of neglect and historical inertia. But they may also be unclear or contradictory because they accurately reflect the contradictory impulses and orientations of the society the agencies serve. Schools attempt to instruct, but also inculcate attitudes toward social behavior and citizenship, not because educators are fuzzy but because these objectives are both favored by the clienteles of schools (and because the case that they are mutually incompatible has not and probably cannot be made convincingly). Criminal justice institutions are oriented toward punishment and rehabilitation not because judges and corrections officials are simple-minded, but because the society has impulses toward reforming as well as deterring criminals.

The public service areas of education, corrections, and welfare in recent years have all been subject to efforts to increase accountability through goal clarification. Educators have wanted to concentrate on reading to the exclusion of other educational objectives; corrections analysts have sought to clarify the role of punishment and make it more certain at the expense of emphasis on rehabilitation; welfare reformers have successfully separated decisions on income support from social service provision. The dilemma for accountability is to know when goal clarification is desirable because continued ambivalence and contradiction is unproductive, and when it will result in a reduction in the scope and mission of public services. The problem of goal ambiguity has contributed to the discrediting of services in social work, corrections, youth offenders programs, and mental health, and to the dismantling of many programs to provide assistance to those who seek it in these areas. But it requires the most serious inquiry to determine the long-term implications of requiring the clients of these institutions and agencies to have recourse exclusively to community and personal resources.

PERFORMANCE MEASURES

The development of performance measures is critical to a bureaucratic accountability policy. Without knowing how to measure performance, organizations cannot hold employees accountable for the performance. For this reason administrators expend considerable resources attempting to develop performance measures in order to control employees' behavior.

There is no question that public services can be enhanced when valid performance measures are developed. In such cases public service workers can be held accountable for producing results in the same way that machine operators can be charged with producing a certain volume of output in a given period of time. However, public service workers, like machine operators, must also be assessed for quality control, since producing a volume of items is meaningless without consideration of the standard maintained in production. Here, paradoxically, the search for performance measures can interfere with the quality of public service.

In theory, quantitative measures of performance should be fairly easy to obtain, and consent on their validity reasonably uncontroversial. This is not always the case in street-level bureaucracies, however, for several reasons.

First, street-level bureaucrats will concentrate on the activities measured. If police officers are assessed on traffic ticketing or vice arrests, activity in these areas will increase. This is entirely predictable when we recognize that police have control over their search activities, and can choose to concentrate on one dimension of their job or another. By virtue of simply putting attention on some tasks over others, street-level bureaucrats can improve their performance on most quantitative measures managers introduce. If welfare workers are assessed on their error rate, the error rate will go down because workers pay more attention to it. If teachers are assessed on the proportion of their charges who pass year-end examinations, more will pass as teachers "teach the test." This is neither surprising nor in itself deplorable, but simply highly probable. Whenever management undertakes to concentrate on measuring a dimension of performance, workers correctly accept this as a signal of management priority. A problem is created when the measure induces workers to reduce attention to other aspects of their jobs, and when there is no control on the quality of work produced.[12]

Relatedly, street-level bureaucrats will make choices and exercise discretion by directing their activities in ways which will improve their performance scores. This phenomenon did not begin and end with Peter Blau's classic report of the employment counselors who made greater efforts for easy-to-place clients at the expense of more difficult cases when they began to be assessed in terms of successful placement ratios rather than the caseload they carried.[13] The phenomenon of "creaming" in recruiting for social programs, has similar dynamics. Workers select for their programs clients who are likely to do well in them, in order to improve the appearance of success. As James D. Thompson has put it, "Where work loads exceed capacity and the individual has options, he is tempted to select tasks which promise to enhance his scores on assessment criteria.[14] This generalization obtains for individuals and also for the work units of which they are a part.

Fraud and deception can also intrude into performance measurement. The Washington, D.C. police were quite proud of their record of reducing serious crimes until a study revealed that police officers were inexplicably reporting that most burglaries involved items valued at less than $50. Significantly, the definition of a felony for this crime is defined in part as involving the theft of over $50 in value.[15] The incentives to underreport the value of items in burglaries are the same as those which induced New York City sanitation men to water their garbage so that their trucks would weigh the expected amount when they appeared at the landfill site, even though the drivers had not completed their runs.

It may be claimed that these problems–of inducing behavior to conform to the measure, neglecting other responsibilities, and unauthentically performing according to the measured standards–are simply difficulties that skilled management experts can overcome. In particular, management often seeks measures of resource deployment, depending on the inference that the provision of resources is a surrogate for (and, to be sure, a prerequisite to) service delivery. This inference is acceptable when the qualitative issue is resource deployment, as in the case of police dispatch, ambulance response time, and neighborhood shift allocations in sanitation services.[16]

The difficulty arises in the inferences that resource deployment of a particular sort bear a relationship to the quality of service delivery. For example, caseload activity might be used as a quantitative measure of performance, since it indicates formal relationships between street-level bureaucrats and clients. Class size indicates associations between teachers and children. Court dispositions indicate relationships between defendants and judicial personnel. But in all these instances there may be inverse relationships between the quality of street-level bureaucrats' involvement and the number of clients they process. If simply having people processed, or having them attached to public service workers were the issue, these measures would bear a meaningful relationship to desired service. But our expectations of these public services are different. It is not sufficient that people are assigned a social worker, sit in a classroom, or have their cases heard. We also expect that they will be processed with a degree of care, with attention to their circumstances and potential. Thus there may be no relationship, or an inverse relationship, between quantitative indicators of service and service quality.

The more discretion is part of the bureaucratic role, the less one can infer that quantitative indicators bear relationship to service quality. Even in such an apparently straightforward measure as the number of arrests made by policemen, or the number of people treated in emergency rooms, we have no idea whether the arrests were made with care, or that treatment met appropriate standards. Sophisticated management specialists acknowledge the problems of inferring quality from quantitative indicators.[17] But this does not prevent utilization of quantitative measures as surrogates for service quality and the common practice of ignoring the problems of inference in their utilization.[18]

Of course the reason quantitative measures are used so often is that actual performance is virtually impossible to measure. It is perhaps useful to put this quite bluntly: we cannot measure the quality of street-level bureaucrats' performances, particularly in terms of the most important aspects of their jobs. Aspects of performance can be measured and assessed, and many surrogates for performance measures can be developed with important implications as management tools. But the most important dimensions of service performance elude our calibration.

Measures of performance quality are elusive for reasons analogous to the difficulty in circumscribing street-level bureaucrats' discretion. If clients or presenting cases should be treated as if they might present unique situations, then it is impossible to reduce responses to sets of appropriate and previously indicated reactions. To put it another way, the more street-level bureaucrats are supposed to act with discretion, and the broader the areas of discretionary treatment, the more difficult it is to develop performance measures. If we are not agreed as to what comprises good teaching, how can we measure it? If we are not willing to deprive police officers of discretion because on the street they need to be able to make judgments based upon an appraisal of the total situation, how can we propose measures of quality arrests and interventions with citizens? If every client should be treated as if he or she may require responses tailored to the individual, how can we specify of what a good interview consists?

It may be argued that we may still assess service quality by developing outcome indicators. But here similar questions arise. First, service quality measures are meaningless without adequate controls to assess levels of difficulty. The same outcome may have required radically different service because of the difficulties presented. For example, the same student achievement levels might represent excellent work on the part of a teacher of students with learning difficulties, and poor work on the part of a teacher with bright and motivated students.

Without controls there can be no comparability of units of analysis, unless the often unwarranted assumption is made that levels of difficulty are equal. Thus teachers resist being measured by the progress of their pupils unless adequate provision is made to control for their students' previous levels of achievement (and, more importantly) for their students' capacity to learn. Thus police officers would object to utilizing arrests per capita per available officer as a measure of performance unless controls were introduced for the propensity of criminal behavior in the district. Comparing districts by outcome measures tends to be useless because of the inadequacy of such controls.

Some advocate that measures such as these be deployed in order to discover deviations from normal practice, so that measures to workers who deviate from the norm can be brought into line. Here the problem is that unless one is confident that the best workers or districts are doing a good job, such comparisons may simply institutionalize mediocrity.

Street-level bureaucrats' interactions with clients tend to take place in private or outside the scrutiny of supervisors. Interviews are held in private offices and/or under norms of confidentiality. Teaching is done in classrooms which principals and supervisors do not normally enter, and if they do, provide notice, so that the teaching, like a performance, may be changed by the presence of an audience. Police officers, although taking action in public, normally do so in the absence of the observations of other officers or supervisors. The exception is the officer's partner, who is compelled by police norms to shield his partner from criticism. Of the street-level bureaucrats we have studied, only judges tend to make their important decisions in public.

This fact provides a barrier to performance measurement, for it tends to reduce the viability of an important potential source of performance measurement. It might be possible for street-level bureaucrats to scrutinize each others' work and provide assessments of quality. But, given the structure of these agencies, such scrutiny would be highly obtrusive in relations between workers and clients, and very costly if engaged in on a widespread basis. Thus public service agencies rarely engage in direct observation of their line workers, but depend upon the written record supplied by their workers. (The reliability of such records is discussed above.)

ACCOUNTABILITY AND PRODUCTIVITY

Thus far I have focused discussion on some of the major difficulties in developing an administrative accountability policy. But are there any negative effects of such policies? For example, what is the harm of attempting to develop performance measures? It may be difficult to measure performance, but perhaps we are simply at the beginning of the development of a management tool. Perhaps the current measures of performance are not entirely adequate, but they may have their uses, and they may be increasingly refined.

This is the rhetoric of those who are committed to achieving bureaucratic accountability, recognize the inadequacy of current measures, but apparently have faith that their approach is ultimately correct.[19] This line of discussion would have us believe that there are some benefits to current efforts to develop accountability through improved performance measurement, that these benefits are likely to increase, and that there are no significant costs. Surely management benefits from operationalizing and attempting to develop measures of worker performance. Even if the preferred behavior cannot be adequately measured performance measurement and monitoring can signal workers powerfully concerning which aspects of performance are most salient. However, in the current period bureaucratic accountability policies also have negative consequences because of the competing demands on, and of, administrators.

In the current period public agencies are under enormous pressures to minimize costs and increase productivity. They are under pressure to reduce government expenditures or keep them from rising. They are under pressure to increase productivity in order to maintain, or claim that they are maintaining, services in the face of financial stringency. And they are under pressure to increase productivity in order to justify employee pay increases, which they are under pressure to grant. (The only way to stabilize government budgets when services cannot be reduced beyond a certain level and costs are rising is to increase the productivity of the present work force. Organized workers argue that they have no incentive to increase productivity unless they can share in the gains made because they work harder and cooperate with the reorganization of work often entailed by productivity reforms.)[20]

Productivity in the public sector summarizes the relationship between the utilization of resources and the resulting "product" in providing public services. Productivity may improve when costs remain the same while public services increase, when costs decline while services remain the same, or when costs increase but services increase still more. Schematically, there are two dimensions to public services implied here—one qualitative, the other quantitative. If the quantity increases or remains the same, but services have declined qualitatively, productivity increases have not actually taken place.[21] If more garbage is picked up on the streets by the same crews, but half is strewn on the streets, productivity has not increased. The debasement of service is what infuriated New Yorkers recently when transit workers were given an increase in pay based upon gains in productivity, but it then came out that the Transit Authority had been able to provide services with fewer personnel only by increasing the time between trains and reducing the number of cars in operation.[22] In this view, the transit workers had been falsely credited with improved productivity.

These are the essential elements of productivity. In practice, however, debasement of services is rarely taken into account in productivity practices, although the problem is given lip-service by productivity theorists. First of all, if the quality of service is difficult to measure, so is reduction in service quality. Second, there are many ways to save money by eroding the quality of service without appearing to do so. They include offering services on a group rather than individual basis, substituting paraprofessionals (often paid from other sources) for regular staff, and, conversely, forcing professionals to handle clerical and other routine chores which reduce the time they have to interact with their clients.[23] Additionally, street-level bureaucrats can narrow the range of situations in which they will act. Examples include a legal services office which decides to take only emergency cases, police departments which decide to neglect selected infractions, and schools which offer a reduced program of learning opportunities. Each of these techniques permit managers to give the appearance of maintaining services while reducing costs.

Third, in the current period, pressures experienced by public managers to reduce the budget and improve productivity are pressing and general, while the constituency for maintaining service quality is disorganized, quite weak, or nonexistent. Only clients experience service quality reduction, and they are severely constrained in comparing their experiences with others, and organizing collectively to oppose service quality debasement. Ironically, the greatest opponents of service quality debasement are street-level bureaucrats themselves, for whom debasement often means harder work, less job satisfaction, and greater individual problems with clients. Yet they are cross-pressured by the interest they have in helping their agencies appear financially responsible, and their collusion with public officials to share financially in productivity increases.

But there is more to the debasement of public services than pressures and interests. A large part of the problem stems from the orientation toward measurement, precision, and scientific management itself. Consider the formula *productivity = service quantity and quality/cost.* Two of the terms, service quantity, and cost, are easy to measure; the third is virtually impossible. Managers under pressure to improve productivity are likely to try to cut personnel or obtain more work from existing personnel because these are the terms of the equation for which measures are available and which managers can manipulate. Thus staffs are reduced to bare bones without reduction in responsibilities. Thus staffs are asked to do more without increases in personnel.

STREET-LEVEL BUREAUCRATS AND THE FISCAL CRISIS

There is always an implicit tension between resource constraints and the inexorable demands for public service. However, ambitions to assess bureaucratic priorities every year to the contrary notwithstanding, this tension is rarely manifest politically. The budgets (and employment rolls) of street-level bureaucracies rise not only with increases in population to be served but also with higher standards of what citizens are entitled to, and decline in the availability of comparable private services and the perceived need for more effective and improved agencies of social control–coercive or manipulative. The impulses to increase expenditures in these areas are rarely challenged in terms of resource constraints. At least this has generally been the case since World War II, aided by federal government subsidies to state and local public services, which have postponed or softened the confrontation between revenues and expenditures in areas such as public health, education, police, and welfare. However, the current period, characterized as a "fiscal crisis" among state and local governments, forces recognition of the relationship between what people get from government and what jurisdictions are willing to pay.

At best the term fiscal crisis is reserved for situations in which financial agreements and long-standing patterns of practice can no longer be honored, as when a political jurisdiction cannot meet its payroll or honor its commitments to lenders. But the term is also used much more loosely to mobilize people to believe that there is something wrong or there is a problem associated with current and projected expenditures relative to available revenues and other income. If political and economic elites are successful in promulgating a sense of crisis, they are able to make manifest and set the terms of confrontation between governmental expenditures and income. If in other times social services (for example) grew in response to perceived societal needs, in a fiscal crisis the imperatives for service development are subordinated to the demands of perceived revenue limitations.

Like other political confrontations, the management of fiscal crises have redistributive consequences. The costs of responding to the needs of expenditure constraints do not fall evenly or randomly on the population as a whole, but rather affect different segments of the population differentially. The fiscal crisis of the cities provides a focus and an apparently benign rationale for attacking and injuring the provision of public services. And they demonstrate the vulnerability to attack of maintaining high levels of public service quality.

The "fiscal crisis" of the cities affects the quality of service delivery in two significant ways. First, services are rationed in various ways, maintaining the appearance of service while reducing and debasing it in practice. This is not to say that legitimate savings cannot be realized by eliminating "real" waste and duplication that may have existed.[24] However, in city administration these "real" savings tend to be concentrated in areas in which questions of resource deployment are paramount, not in areas in which the provision and nurturing of interactions with street-level bureaucrats is at issue. It is conspicuous, for example, that when administrators take pride in productivity savings, it is in the sanitation department that the greatest successes are often realized. In police departments dispatch (deployment)–not interactions with citizens–is the area of concentration in productivity campaigns. Most street-level bureaucracies have to contend with impulses to reduce the amount of time workers spend with clients, not the reverse.

For public officials the problem of managing the fiscal crisis consists of reducing expenditures while minimizing the apparent impact of the cuts. This is why initially cuts will be said to eliminate waste and duplication whether or not waste and duplication indeed were taking place. Rationing typically means increasing the costs to clients of seeking services, while maintaining the service shell, or reducing services to decrease potential benefits. Both prospects are likely to lead to lower client demand. Closing neighborhood branch offices while continuing to offer services from a central (downtown) location is a typical technique for achieving this. Reducing the number of telephones or receptionists reduces the number of inquiries. Increasing the response time in investigating a complaint reduces the efficacy of complaining and hence future volume. City agencies often experience decreasing ability to process citizen needs at the same time that they are experimenting with techniques to improve their performance. The public message regarding agency responsiveness becomes mixed, to say the least.

When public managers decide to fight demands for service reduction they say that all waste, duplication, and nonessential services have been cut, and that any further cuts will be in essential services. Their ability to make this claim depends upon general public perception of the importance of the agency and public employees' collective capacity to resist. Thus schools can make the claim more effectively than welfare agencies, police departments more effectively than sanitation departments. Cuts in service provision obviously may affect service quality, but it is impossible to determine from public rhetoric where the politics of distributing urban resources to public employees ends and injury to service delivery begins.

A second dimension of response to the fiscal crisis is personnel reductions. Personnel practices are particularly important because salaries comprise the bulk of urban budgets; thus savings must be sought in the area of public employment.

Personnel practices in the fiscal crisis tend to follow a path of severity. Administrators first make it more difficult for agencies to replace workers who leave. Next administrators suspend hiring, then slow wage increases or freeze wages, and then begin to lay off workers. All of these steps have implications for service provision, but the most important point is that each of these steps reflects increasing penalties to public employees, regardless of the implications for clients. Priorities are set by imperatives of labor-management relations, not by the needs of clients for service provision.

Increasing the difficulty of replacing workers and freezing employee rolls, represent public managers' efforts to realize savings through attrition. They attempt to reduce personnel rolls by not replacing those people who retire or otherwise leave their jobs. Since the rate of exit in almost any line of work in normal circumstances is substantial, a significant reduction in employment can be realized over several years without firing anyone.[25] Realizing savings through attrition accommodates public managers' needs to keep peace with organized public employees, but has substantial costs for service provision above and beyond the obvious reduction in force. Since workers do not retire or otherwise leave their jobs in response to agency priorities, the incidence of turnover is unevenly distributed in the work force. This means that important gaps in service provision occur. Workers who are important to the operation of a particular office or who possess critical skills often will leave the work force rather than the most marginal employees. If the critical positions they vacate are left unfilled, the injury to service provision is obvious. But even if they are filled by employees who remain, they are unlikely to be filled well.

In the street-level bureaucracies with a high level of job specialization the vacancies will be filled by employees who lack the required skills or resources. Fifth-grade teachers will be assigned to kindergarten classes, and physical education instructors will become math teachers when there are excesses of the former and need for the latter. But even in street-level bureaucracies with low levels of specialization it will be difficult to fill the vacancies adequately.

Police departments, for example, assign desk officers to active patrol in order to maintain the patrol force when additional officers cannot be hired. But those officers assigned to desk jobs are likely to be more suitable for sedentary positions than the officers they replace. Moreover, except for retirees, the people who leave public service work in the fiscal crisis will tend to be the more employable, so that the workforce experiences a mediocritization through attrition at the same time.

Street-level bureaucrats' performance is not tied directly to wage incentives. Promotions and raises, when they are given out, do not depend so much upon job performance as on personal relations, additional outside training, workload "handling," and other factors unrelated to client servicing. Moreover, promotions to positions of greater responsibility are rare, since the job hierarchies in most street-level bureaucracies schematically resemble relatively flat pyramids, with jobs existing undifferentiated at the bottom of the scale. When most teachers can look forward only to being teachers, most police officers only to being police officers, incentives to high-quality service may flag unless specially encouraged.[26]

For these reasons it is incorrect to argue that wage freezes directly affect workers' motivation. Their impact is somewhat different. First, wage freezes and slowdowns affect the likelihood that workers will stay in their jobs and force people out of jobs without firing them. Again, this affects the distribution of age and skills in street-level bureaucracies. Older workers will stay to protect and build pension benefits. Younger workers will be more likely to leave public employment, somewhat tempered by the availability of better jobs elsewhere.

In periods of high inflation, wage freezes effectively reduce workers' real income as well as their income relative to workers in other sectors. The feelings of deprivation that may result from diminishing workers' income in these ways are quite different from attitudes toward wages which street-level bureaucrats may have had in other periods. While wages may not play a major role in creating incentives to performance, effectively reducing wages may have a significant impact on street-level bureaucrats' attitudes toward their jobs.

Street-level bureaucrats may be accountable to managers and clients, as mentioned above. But they are accountable to clients *on behalf of their agencies and the public purposes they represent.* To reduce workers' wages is to shred these bonds of accountability by bringing to the surface a reality that otherwise remains obscured in street-level bureaucracies by the claims and ideology of professional status and attitudes. This reality is the wage relationship between street-level bureaucrats and their agencies. People who accept the relatively fixed civil-service formulas for wage increases will not easily accept receiving less than they did receive, particularly as the motivation for street-level work leads some employees to regard their work as voluntaristic to some degree, that is, undercompensated to begin with. As wages are effectively reduced, street-level bureaucrats may be expected to look more to their benefits and remuneration, and less to the service dimensions of their job.

Decruitment practices, of which firing workers is the most drastic, have similar implications.[27] When workers are not replaced, their responsibilities are distributed among those who remain, usually without reducing the responsibilities they already have. Increasing the possibilities of other staff without increased compensation or work resources is the white-collar equivalent to an assembly-line speed-up, than which there is nothing worse in the life of an industrial worker.

In this connection, consider the modest complaint of a New York City school teacher to staff reductions, increased responsibilities, and reduced resources.

> I have never been quite sure what ["increased productivity"] means exactly. However, if it means what I think it means, they would like to see us work harder than we ever did before. If this is so, then all the proponents of "increased productivity" will be delighted to know that we are doing remarkably well in that department.
>
> For example, we have official classes of 45 or more youngsters and ten minutes in which to take attendance, read circulars, distribute notices, make reports (in duplicate yet), answer questions, etc. In many cases we have classes which have rosters of 49 or more children with 30 chairs in the room or typing classes of 47 with 32 typewriters. Add to this emergency coverages of classes, cafeteria patrol or other building assignments, program problems, shortage of supplies and equipment and much more—all of this with reduced staff—not to mention the mounds of work we take home with us. The pressures under which we work can never be understood by anyone who is not involved in day-to-day school activity. Yes, we have indeed increased productivity, but in so doing we have decreased our effectiveness as human beings to our students, our families and ourselves.[28]

When workers are fired managers also abridge the implicit contract with employees, by bringing to the surface the reality of job insecurity in what previously had seemed secure employment. Again, those who remain carry the workload of those who were fired. While some increased and heroic efforts may be forthcoming, it is equally likely that work will be processed in ways which reduce the amount and quality of time street-level bureaucrats can spend with clients.[29]

Again, this *de*cruitment does not take place according to calculations of impact on service, but in response to the ethics of seniority. For example, different public agencies are typically asked to reduce their forces by a certain fixed percentage. Politicians do not want to choose between services and so establish a decision rule for work-force reduction. Street-level bureaucracies which do choose between units or services undermine the strategic implications of their choices by providing those with seniority the chance to "bump" less senior employees from their positions in sectors of the service which are untouched by the budget cuts.

For some enterprises reductions in personnel rolls can be cleansing, but public agencies cannot take advantage of opportunities to eliminate ineffective workers. Particularly affected by derecruitment will be younger workers and those who have been recently hired, often members of minority groups. It is difficult to specify the impact of decruitment precisely, but we can see that it falls unevenly on those with more recent training and new ideas, and those with fresh perspectives. It falls less heavily on those with more experience–an arguable benefit.

These observations concerning reductions in workforce should not lead us to conclude that street-level bureaucracies should be the size they are, or larger, or that they should never be smaller than they presently are. The fact is that we know very little about the "proper" size of street-level service delivery units. A public school deprived of its specialists might actually serve students better by throwing regular teachers upon their own resources. The important point is that when public bureaucracies are growing, jobs usually are created in response to perceived needs, with workers hired with reference to credentials and apparent qualifications. But when they are forced to shrink, public bureaucracies rarely remove workers in response to decisions concerning the most effective utilization of reduced available resources.

Who speaks for service quality in the era of performance measurement, productivity campaigns, and fiscal crisis? Public managers, with better control over costs and resource deployment than over the quality of the "product," sacrifice service quality in the name of efficiency and productivity. Street-level bureaucrats more and more are reduced to production units whose work is speeded up and whose managers appear content to sacrifice quality in order to maintain volume. In the process the conditions of work are eroded and workers are unable to utilize many of the coping mechanisms and attitudes which helped sustain their jobs under difficult conditions in earlier periods. Thus, the fiscal crisis raises the salience of the wage relationship and diminishes the salience of service. This is ironic since the wage and benefit demands of organized public employees have been widely regarded as one of the primary causes of the fiscal crisis in the first place.

This paper has drawn attention to the contributions of bureaucratic accountability policies to exacerbating problems of the quality of service delivery. But there is more. Such policies set the stage for future management of service delivery, and future conceptions of the role of human services. If current administrative practices erode workers' sense of responsibility for clients, then establishing nonmanipulative, responsive worker/client relationships will be that much harder to establish in the future. When qualitative aspects of service delivery are neglected, cost reductions and volume receive more attention as workers and managers accommodate their behavior to agency signals of priorities. This contributes to the self-fulfilling prophecy of the ineffectiveness and ultimate irrelevance of social services, even though the human needs for nurturing, protection, support, and assistance remain unanswered. Thus the tones of the fiscal crisis may linger even if the budgetary alarms of the current period are eventually quieted.

NOTES

1. See Edward Wynne (1976: 30-37).

2. This analysis will be elaborated in an extended treatment of the work structure of street-level bureaucrats and their role in public policy, now in preparation. For a preliminary statement see Michael Lipsky (1976: 196-213). Also see Richard Weatherley and Michael Lipsky (1977).

3. On coping strategies of lower level workers see Chris Argyris (1964: ch. 4). Generally see Richard Lazarus (1966).

4. On clients' ability to discipline organizations by withdrawing patronage, see Amitai Etzioni (1958: 251-264; Albert Hirschman, 1977).

5. See David L. Kirp (1973: 705-797); Ray C. Rist (1975: 517-539).

6. A very significant portion of state and local government expenditures goes to pay the salaries of street-level bureaucrats. Public school employees, for example (3.7 million people in 1973), represent more than half of all workers employed by local governments. More than five out of seven of these are instructional personnel. Of the public employees not engaged in education, approximately 14% work in police departments. Approximately 54% of all local government payroll expenditures went to public education in 1973, almost 80% of which went to instructional personnel. Police salaries comprised one-sixth of local public salaries not assigned to education. Derived from Alan Baker and Barbara Grouby (1975: 109-112, Table 4/3).

7. This is not the case with all "buffer" roles, played by people who represent organizations to the public. For example, salesmen are not expected to be responsible to buyers in anything like the same sense that, say, social workers are expected to be responsible to clients. See the discussion of buffer roles in James D. Thompson (1973: 191-211).

8. I am not arguing that discretion never can and should be reduced. On the contrary, where lower-level workers usurp discretionary powers it is obviously appropriate for management to intervene. (For an example of such usurpation, see Irwin Deutscher (1968: 38-52). However, when instances of appropriately circumscribed discretion are exhausted the basic work of street-level bureaucrats remains.

9. For a discussion of the problems of record-keeping and accountability in medicine, see Eliot Friedson (1976: 286-298).

10. The best discussion of the effects of weak management sanctions on developing

norms of reciprocity supportive of low levels of effectiveness is Eric Nordlinger (1972: ch. 3).

11. Murray Edelman discusses the symbolic implications of administration and bureaucracy for mass democracy in *The Symbolic Uses of Politics* (1964: ch. 3).

12. James Q. Wilson (1968: 291) describes this tendency for police departments. "The police supervisor . . . would have to judge his patrolmen on the basis of their ability to keep the peace on the beat, and this . . . is necessarily subjective and dependent on close observations and personal familiarity. Those departments that evaluate officers by 'objective' measures (arrests and traffic tickets) work against this ideal."

13. See Peter Blau, *The Dynamics of Bureaucracy* (1963: 36-56).

14. See James D. Thompson, *Organizations in Action* (1967: 123).

15. See David Seidman and Michael Couzens (1972). Perhaps because they are subject to considerable scrutiny, illustrations of manipulation of statistics by the police are more likely to come to public attention than other public service agencies. See, for example, the criticism of an experiment in Orange County, California, which provided incentive pay increases to police officers for crime reduction. A report on this experiment alluded to the "possibility that the increase in larceny represents a shifting of criminal activities or a reclassification of burglaries into a closely related category which will not harm prospects for an incentive reward" (New York Times, 1974, p. 77; also May 12, 1972, p. 1).

16. Significantly, the literature on productivity in public service provision draws its most persuasive examples from these and similar cases of resource deployment. See, for example, Edward K. Hamilton (1972: 784-795).

17. A good discussion of the problems of inference is found, for example, in Harry Hatry (1972: 776-784).

18. Hamilton, previously cited (1972: 787), specifically commends the utilization of quantitative measures "where output is very hard to measure . . . to improve the deployment of resources so as to maximize the probability that our resources will be available at the time and place they are needed most." This may be useful for fire protection where the *presence* of fire fighters is the critical aspect of service provision. But it cannot be adequate for street-level bureaucracies, when resource availability may not be related to service quality.

19. Consider the following paragraph:

> Admittedly, there is an unevenness to productivity measurement. Some measures are relatively sophisticated, others crude. But in the common absence of any yardstick of productivity, even crude information is of value. At least it is a means of introducing systematic quantitative analysis into the decision-making process. Once that precedent is established, incremental refinements will undoubtedly lead to more sophisticated measures. Quantifications should only be attempted, however, if the organization has the qualitative and technical capacity to interpret and apply data meaningfully. [Holzer, 1976: 19]

20. If pay increases for workers and the cost of city services depend upon productivity, then productivity measurement and assessment obviously become highly political phenomena. For example, New York City workers seek to measure the size of productivity savings in terms of the net savings to the city from higher worker output. Fiscal managers, however, argue that productivity savings should be assessed in terms only of lower salary costs resulting from the need for a smaller workforce to accomplish the job. (See the New York Times, March 26, 1977.)

21. For a discussion of these elementary aspects of productivity see, e.g., Nancy S. Hayward (1976: 544-550).

22. See the New York Times (1976b).

23. For a discussion of some of these service rationing practices see Weatherley and Lipsky, previously cited.

24. I say "real" in quotes because when a saving is real and when it represents a reduction in governmental effort is an empirical and normative question. Sometimes "crisis" can force management to attend to costs so that real savings are discovered, e.g., energy conservation by reducing unnecessary wattage in bulbs. But at other times a change is simply justified by calling it duplication or waste reduction, although it may not be.

25. The rate of attrition will partly depend upon employees finding other jobs. To the extent that a fiscal crisis is perceived by many potential employers at the same time, causing them to stop hiring, an attrition policy will be correspondingly less effective.

26. In part, promotion and retention in street-level bureaucracies are not based upon the quality of service provision because service provision is so difficult to measure. Hence, surrogates for effective service provision—such as tenure and advanced training—often bearing little relationship to worker effectiveness, are used extensively to reward and promote workers. These are not generally contradicted by more appropriate service delivery measures. On promotion in street-level bureaucracies see, for example, John Van Maanen (1973: 54); David Goodwin (1977: 66-67).

27. The phrase is drawn from Donald H. Sweet, Decruitment: A Guide for Managers (1975).

28. Letter of Hanna B. Leibowitz (1976).

29. Perhaps the most neglected aspect of the fiscal crisis is the extent to which the firing of public employees represents a reduction in one of the critical functions of big city governments—the provision of relatively secure and decent jobs. After spilling many words on the fiscal crisis the New York Times (1976a) editorialized recently as follows: "The trouble is, the bureaucracy also consists of people. Thus the fiscally sound demand for greater economy and efficiency in the municipal health care bureaucracy could lead to the discharge of thousands of hospital workers. In the absence of alternative job opportunities, the result would be suffering and despair in minority communities—and a sharp increase in welfare rolls." Frances F. Piven (1976: 8) has written persuasively on the redistributive aspects of urban fiscal liberalism and stringency.

REFERENCES

ARGYRIS, C. (1964). Integrating the organization and the individual. New York: John Wiley.

BAKER, A., and GROUBY, B. (1975). "Employment and payrolls of state and local governments, by function: October, 1973. Pp. 109-112, Table 4/3 in Municipal year book, 1975. Washington, D.C.: International City Management Association.

BLAU, P. (1963). The dynamics of bureaucracy (rev. ed.). Chicago: University of Chicago Press.

DEUTSCHER, I. (1968). "The gatekeeper in public housing." Pp. 38-52 in I. Deutscher and E. J. Thompson (eds.), Among the people: Encounters with the poor. New York: Basic Books.

EDELMAN, M. (1964). The symbolic uses of politics. Urbana, Ill.: University of Illinois Press.

ETZIONI, A. (1958). "Administration and the consumer." Administrative Science Quarterly, 3, 2(September):251-264.

FRIEDSON, E. (1976). "The development of administrative accountability in health services." American Behavioral Scientist, 19, 3(January/February):286-298.

GOODWIN, D. (1977) Delivering education services: Urban schools and schooling policy. New York: Teachers College Press.

HAMILTON, E. K. (1972). "Productivity: The New York City approach." Public Administration Review (November/December):784-795.

HATRY, H. (1972). "Issues in productivity measurement for local governments." Public Administration Review (November/December):776-784.

HAYWARD, N. S. (1976). "The productivity challenge." Public Administration Review (September/October):544-550.

HIRSCHMAN, A. (1977). Exit, voice and loyalty. Cambridge, Mass.: Harvard University Press.

HOLZER, M. (1976). Productivity in public organizations. Port Washington, N.Y.: Kennikat Press.

KIRP, D. L. (1973). "Schools as sorters: The Constitutional implications of student classification. University of Pennsylvania Law Review, 121(April):705-797.

LAZARUS, R. (1966). Psychological stress and the coping process. New York: McGraw-Hill.

LEIBOWITZ, H. B. (1976). Letter of Hanna B. Leibowitz. New York Times, September 28, p. 38.

LIPSKY, M. (1976). "Toward a theory of street-level bureaucracy." Pp. 196-213 in W. Hawley and M. Lipsky (eds.), Theoretical perspectives on urban politics. Englewood Cliffs, N.J.: Prentice-Hall.

New York Times (1977). March 26.

——— (1976a). November 9, p. 36.

——— (1976b). October 22, p. A26.

——— (1974). November 10, p. 77.

——— (1972). May 12, p. 1.

NORDLINGER, E. (1972). Decentralizing the city: A study of Boston's little city halls. Cambridge, Mass.: M.I.T. Press.

PIVEN, F. F. (1976). Boston Globe, December 9, 1976, p. 8.

RIST, R. C. (1975). "Student social class and teacher expectations: The self-fulfilling prophecy in ghetto education." Pp. 517-539 in Y. Hasenfeld and R. A. English (eds.), Human service organizations. Ann Arbor, Mich.: University of Michigan Press.

SEIDMAN, D., and COUZENS, M. (1972). "Crime, crime statistics, and the great American anti-crime crusade: Police misreporting of crime and political pressures." Paper presented at the annual meeting of the American Political Science Association, Washington, D.C.

SWEET, D. H. (1975). Decruitment: A guide for managers. Reading, Mass.: Addison Wesley.

THOMPSON, J. D. (1973). "Organizations and Output transactions." Pp. 191-211 in E. Katz and B. Danet (eds.), Bureaucracy and the public. New York: Basic Books.

——— (1967). Organizations in action. New York: McGraw-Hill.

WEATHERLY, R., and LIPSKY, M. (1977). "Street-level bureaucracy and institutional innovation: Implementing special education reform." Harvard Education Review, 47, 2(May):171-197.

WILSON, J. Q. (1968). Varieties of police behavior. Cambridge, Mass.: Harvard University Press.

WYNNE, E. (1976). "Accountable to whom?" Society, 13, 2(January/February): 30-37.

2

Professional Self-Regulation in the Public Interest: The Intellectual Politics of PSRO

SCOTT GREER

☐ IN OCTOBER, 1972, the Congress enacted legislation requiring that all health care delivered under Medicare and Medicaid be monitored and continuously evaluated. The device specified in the bill was peer review, e.g., selected groups of physicians are to be held responsible for the quality and cost of medical services, beginning with institutional treatment, but presumably to be extended to private practice. While these Professional Standards Review Organizations operate at the local level, and are usually made up of local physicians, if the local groups are unable or unwilling to form such an organization, they will be appointed by the Secretary for Health, Education and Welfare, from elsewhere if necessary.

The reasons adduced for this action were uncertainty over doctors' willingness or ability to police themselves. It was thought that patients were frequently overdoctored, overdrugged, and overhospitalized; that fees were sometimes outrageous; that some doctors were doing work they were not qualified to do, and some were doing sloppy work. It was known that costs were skyrocketing, e.g., Medicare premiums doubled between 1966 and 1972. It was thought that the PSROs would improve the quality of medical care and help to contain costs.

The Secretary designates regional jurisdictions for PSROs, so choosing them as to allow for economical use of mass data analysis. (This would

---

AUTHOR'S NOTE: *This is a revised version of a paper presented at the Conference on Professional Self-Regulation, Public Health Service, June, 1975, and appeared in the Proceedings.*

ordinarily require at least 300 physicians.) Existing societies which represent the physicians would have first bid to establish Organizations provided they were open to all qualified physicians and were representative (and special care will be taken to ascertain representativeness through periodic voting.)

PSROs will consist of a council, an advisory board, and whatever staff is necessary. The council members must be physicians, and only those who have the privilege of practicing in local hospitals are eligible to sit on cases involving hospital care. The advisory board is made up of health professionals other than M.D.'s or Drs. of Osteopathy. Above the local level there will be State Councils of PSROs, and a National Council to top it off. Each higher level council will coordinate the lower ones, disseminate information, and review their work effectiveness. All expenses will be borne by the Office of the Secretary.

In evaluating the necessity and quality of care, the Organization will develop extensive data banks for practitioners and other providers. These will be used for preliminary screening through computers, monitored by paramedical personnel, and will provide what are called "profiles" of patients and practitioners. Extreme variation from the statistical norm will result in evaluation by the council of physicians.

The sanctions for providers who refuse to mend their ways after their deviation is made known to them are several. As is the case now, Medicare and Medicaid remuneration may be refused for work completed; in addition, the Secretary can terminate or suspend payment to the provider, or assess an amount "reasonably related to the cost involved" up to $5,000. There are, of course, elaborate mechanisms for appeal including, as a last resort, the courts.

This is a radical program of action for a society which has given its medical practitioners almost total freedom in the past. It is a specifically American invention, a typical answer to a problem in social control–professionals will regulate themselves in the interest of public accountability and the public. This came about because (1) regulatory devices from the past no longer worked to control prices and quality of care and (2) even more important, *trust* in those devices was rapidly eroding. Before examining the causes, and PSROs as possible cures, let us first discuss the basis of trust as it was before the "health care crisis."

AUTHORITY AND TRUST IN THE PHYSICIAN

Traditionally, faith in the authority of the healer has been grounded in belief in his personal expertise, a false sense of security, and a degree of that fatalism which remains an important element in the most sophisticated societies. The culture-complex of faith emerged from the community in which the physician had his place in the division of labor. He was related to the patient through common experience and primary relations, grounded in the everyday world of the town; so was he related to his fellow physicians. The average person who bought his services trusted him as a member of the local community, the professional community signified by his diplomas, and the local medical fraternity. As for those who could not afford to buy medical services, they were wards of charity dispensed by outpatient clinics of hospitals or the solo practitioners. They were assumed to have no basic rights to such services and, in a class system based upon massive repression of demand among the poor, were presumably grateful for such care as they received. The use of their bodies as "clinical material" for research and the teaching of fledgling physicians was viewed as no more than a fair exchange.

I have spoken of the "false sense of security" which undergirded the authority of the physician. This sense of security derives from the fact that most people are defined by themselves and others as "healthy" most of the time; further, those in the middle years of life, having survived the perils of birth and childhood and at the peak of their energy, are also most influential in the affairs of the community. The norm is health; from this one assumes the "health services" are being performed. Epidemic illness, bizarre breakdowns and the like occur rarely and can be imputed to a hostile nature, devils, or the will of God; in short, they can be accepted fatalistically because they are culturally so defined.

The authority of the physician is further protected by the norms within the company of equals, his fellow physicians. Eliot Friedson (1972) has demonstrated the strength of these norms, grounded in the common vulnerability of physicians to the problems of a complex and contingent biological world. Such norms were also legally underwritten by the states which generally required that, in the unlikely event a physician was prosecuted for malpractice, testimony against him must reflect the norms of medical men in his own locality. He was judged by his local fraternity. This acted to discourage deviants and to protect the status quo. Further, in shielding physicians from charges of the incompetence and chicanery found in other occupations, the "locality rule" enhanced their status as above the human norm.

This image of the unfailingly competent and benevolent physician was nurtured by his emerging role in the local community. It was, of course, reinforced by organizations which grew out of the physicians' roles and problems; the American Medical Association in particular took itself to represent the entire function of healing and the craft of the doctor. The image became further rationalized and broadcast through the demands of the mass media drama, in a highly differentiated society, for easily recognized characters in situations provocative of suspense and resolution. Thus the role of family doctor which developed in small-town society became a key stereotype in the major communication flow of metropolitan America. It is likely that such tales on television combine with paid advertisements to support and increase demand for the doctor's services.

METROPOLITAN SOCIETY AND THE ORGANIZATION OF MEDICINE

Metropolitan society is very different from the small-town society which nurtured the stereotype. Metropolitan areas are large in population and size; something over 200 of them contain a majority of the wealth, work, and people in the United States. The size of the community, in turn, has resulted in the separation of workplace and market place from the residential area, while the neighborhoods contain widely differing populations. Metropolitan society aggregates populations differing by social class, ethnicity, and life-style. On these grounds types of people are concentrated by residential neighborhoods, segregated from those who differ from them (see Greer, 1962).

Thus the customer's relationship with the merchant is no longer embedded in a community which they share, whether the merchant sells television sets or health care. The relationship is increasingly confined to the specific transaction; that is, it is segmentalized. As for the "local" community of doctors, it is scattered over many miles of city containing hundreds of thousands of patients. The doctors' professional community contains few people from the immediate area of his practice, and many from strange areas of the metropolis; few people in his neighborhood know him personally. His relationship with the patients in "his practice" is not only segmentalized, it is increasingly privatized. The patient can count on neither his commonality with the physician in the local community nor on the fraternity of local physicians as grounds for trust. Authority is increasingly technical in its base: the doctor as replaceable technician, to be judged like any other, is a plausible definition.

So much for those who can afford to purchase medical care, directly or through an insurer. For many others, the older system of charity based upon the physician as Robin Hood no longer exists. The doctor whose rich patients allowed him to give service to the poor no longer practices in neighborhoods where the poor exist, while in the metropolis the concentration and segregation of the poor create vast areas where few (if any) licensed physicians practice. The charity track of the medical system has broken down and so has the moral code which repressed demand from the poor; the result has been the *political demand* for medical care among those who have little effective market demand. The trend is toward increasing organization of medical services for the poor in medical centers and outpatient clinics, large-scale groups financed by the Federal Government. In such groups, the authority of the physician is still accepted by the docile poor but questioned by those who are busy fabricating a culture of client pressure upon public agencies delegated the care of the poor. In such a situation the authority of the physician is ambiguous: is it grounded upon his status as a practitioner of scientific medicine, or as an official of the state?

Authority based upon scientific medicine is not without its constraints. For science is in its essence universalistic, applying to all similar events without respect to geography or patient. Thus the physician is increasingly subject to evaluation in terms that go far beyond his local community, even if such an entity can be defined; the death of the "locality rule" in malpractice trials grows directly out of the rise of universalistic standards for medical practice and malpractice. The image of the physician as competent has suffered considerably, as that competence is evaluated in the courts by experts from outside the local community and medical fraternity.

As for the benevolence in the image of the physician, it has suffered from the singular devotion of the doctors' publicists to the interests of the small businessman, who coexists with the Samaritan and scientist in most practitioners. It becomes increasingly apparent that the physician who diagnoses, prescribes, executes, prices, and judges the final outcome is liable to the charge of "conflict of interest." Note that this charge can be made, whatever the facts of a case, simply because there are no social controls over the transactions between doctor and patient. While this has always been true to some degree, it is increasingly apparent in metropolitan society where the community cohesion of the local area has broken down and the passive submission of the poor is less than dependable. It is clear that in secular society, with segmental relationships between physician and patient, the market nature of their relationship places a great burden on both parties. How can the doctor merit trust? How can the patient know who to trust?

Nor is medical science the firm foundation for practice once anticipated. In an intellectual tradition devoted to the continual analysis of the unknown, a growing area of light multiplies the perimeter of darkness and our ignorance grows faster than our knowledge. In more mundane terms, clinical findings suffer from ambiguity due to the unknown generalizability of most samples: only certain people, of certain types, in certain places seek out given doctors or clinics. In Kerr White's (1975) words, the numerator of the fraction may be known, but the denominator is missing. Thus we have no probability statements running from signs, to symptoms, to underlying condition; from perscription to cure or palliation. Application is weakened by this lack of laws of application, and blind "experiment" by the practitioner becomes likely.

Then too, physicians are increasingly aware of contextual problems in the diagnosis of health or illness. The social situation of the patient, the genetic context, psychological stress, are all candidates for effective causation, potentially more important than the causes he can control. The socially based definition of illness and health, most clear in the case of mental illness, casts still another shadow upon the clarity of clinically based medicine. In short, the illness or health of the patient may easily be due to causes beyond the intellectual or operational reaches of a physician, no matter how sound his clinical training.

Finally, one must reemphasize the dynamic role of technological change, common to large-scale societies. The long-term research and development planning, made possible by large bureaucratic organizations, continually produces new findings about health and illness, new therapies, new tools. The medical doctor is no more immune from technological obsolescence than the computer engineer. Rapid change in theory and practice produce a conscientious imperative to learn and change, while the pressure of daily work severely limits the time and patience the physician has left over for self-education. With these conditions in mind, it is fair to say that the man true to the norms of his formal education as a practitioner of scientific medicine is bound to suffer the results of informational overload. His patients are exposed, as he is, to random and sensationalist reports of new medical developments, and the question of trust is exacerbated. It is little wonder that a substantial minority of physicians lean heavily on paid advertisements by drug corporations in their professional journals for news of innovations, while the primary source of health information among their patients is advertisement on commercial television.

THE DISENCHANTMENT OF THE HEALER

The intellectual concomitants of increasing organizational scale do not cease while growth of the society continues. The disenchantment with the healer who, as Doctor of Philosophy and expert on the gods and devils, fell before the new power of "scientific medicine," has quietly continued its work: Suzanne Langer (1957) has remarked that such disenchantment is not always, or primarily, due to the analytical sciences: more to the point is our increasing knowledge of history. It is demographic and medical history which have been the corrosives attacking the myth of the all-purpose scientific healer who has conquered much and will conquer all.

And the key attack has been on the basic proposition, which stands earnest for the future: just what has scientific medicine conquered? What has the physician in his personal transactions with the individual patient contributed to the health of the human species? A small but influential body of doctrine holds: very little. A careful look at population statistics, death rates, and morbidity rates, leads to the conclusion that most of the triumphs ascribed to scientific medicine predated its existence. Broad trends in societal scale and energy transformation were the real causes of increased life-expectancy and health (McLachlan and McKeown, 1971; McKeown and Lowe, 1966; Sigarist, 1975).

The major dialectic in human history is not, of course, the purely ideational conflict and resolution posited by Hegelians. Nor, on the other hand, can it be subsumed under the conflict between haves and have-nots which Marx and his followers assumed. The broader approach is to see a cultural system prescribing a general course of action, whose consequences in turn falsify the beliefs that led to the act—creating conflict in thought and action, as well as new intellectual and actual opportunities. In short, the dialectic is between action and its results, on one hand, and our interpretation of those results on the other.

The correlation in time of the growth of modern science, the increasing security of doctors in their licensed privilege, and the increase in hospitals and other facilities, with improved probabilities of life and health for human populations in the West, has been interpreted as prima facie evidence for the effectiveness of scientific medicine. The counter-interpretation emphasizes quite different causal factors.

The oldest ground for decreased mortality and improved health among human groups is an *increase in the amount, quality, and dependability of the food supply.* Such a development is often closely geared to increasing societal scale; indeed, the linking of village economies under the authority of a dominant state was a major invention in survival techniques, for it smoothed out the variations in crop success and failure, providing a ready resource in times of famine. The political integration of local areas allows exchange relationships which increase the total amount of food and the probability of adequate distribution (Cottrell, 1955).

As William McNeil remarks in *The Rise of the West* (1970), material artifacts are easier to diffuse among human groups than spiritual concepts, but the easiest items to spread are micro-organisms—bacteria. The involvement of hitherto isolated populations in larger systems, which McNeil calls the "ecumenes," results in new hazards and falling populations. But the populations adjust over time, through the death of those genetically susceptible to the new danger before they reach the age of reproduction. Thus a second major cause for the increase in population and improvement of health in the modern West was *immunization of the population through "natural selection."* Those who were immune survived and reproduced children similarly immune.

The third great cause of improved health and increasd population in the West was the spread of cleanliness among urban populations. The customs which had been selected over longer periods of time for maintaining hygiene in rural villages were not easily carried over into the growing cities of the 18th and 19th centuries. Instead, the cities became hives of concentrated households, bearing various strains of harmful bacteria, while public space often functioned as dumping ground and sewer. The sanitation movement in the 19th century, without a glimmering of the germ theory of disease (indeed, the closest approximation was a theory based upon poisoned air, or "vapors"), nevertheless had powerful therapeutic effects. Its leaders did the right thing for the wrong reasons, cleaning up public space and changing first the norms, then the behavior, of the newly recruited urban population toward *cleanliness of household and person* (Dubos, 1959).

Scientific medicine, in this perspective, had a late and relatively slight effect upon the health of the rapidly urbanizing populations. Some plagues were probably prevented by immunization based upon the germ theory of disease; antibiotics could control actue infections; common casual breakdowns such as diabetes could be treated and arrested, though seldom cured; surgery certainly became safer for the patient as new anodynes were discovered and, more importantly, the principles of cleanliness were applied to the operating room. (Presumably this could have occurred under the doctrines of the sanitationists as easily as under those of Pasteur). Certainly quarantine was of some help to the population at large, but its use had long predated the development of germ theories of disease.

In short, the new critique of the healer emphasizes the societal determination of health and illness, death and survival. Many of the ills of populations were consequences of increasing societal scale, with its concomitants of increased contact with other societies and urbanization, with its effects on the human environment. Many of the cures of such ills also resulted from the further working out of the consequences of our collective actions. The redefinition of health as a public problem, the ability of human populations over time to immunize themselves to strange micro-organisms, and such simple strategies as protecting drinking water and food from rodents had major consequences for human health. The critique implies that the social importance of the personal physician has been inflated out of all proportion to his effectiveness.

Insofar as such arguments have validity, the logically derived priorities for a socially based medicine are as follows. First, a major concern must be with the overall ratio between population and resources, and the distribution of food, clothing, and shelter. Second, within the balance, emphasis must be placed on epidemic control through immunization, and control of the natural hazards consequent to humanity's interaction with nature via developing technologies. Third, such an approach must be concerned with the distribution of medical concern and resources to all of the population, for social and individual benefit.

While the rewriting of medical history has been important in revising the image of the healer, there have been other important forays. These include the development of the "randomized clinical trial," the use of comparative medical statistics, and in general, the movement to apply scientific research methods to the health of whole populations. This movement is based upon a critique of the way medical truth is established and used; it sometimes goes under the name of "epidemiology" (Morris, 1957).

The randomized clinical trial (RCT) is a testing of therapeutic procedures on the basis of a carefully drawn sample of the relevant population (Cochrane, 1972). Its importance is clear when one discovers that most medical research is based upon whatever patients happen to be treated in certain places by certain physicians. RCTs have been further sophisticated through the use of "blind" judgments, in which the physicians making determinations do not know beforehand what they are supposed to find. Such methods have been standbys of scientific research for some time, but such is the isolation of medical research and such the power of the clinical mystique that they have been applied to medicine only recently. The results resemble those of early sample survey research in sociology: the major findings support the "null hypothesis." That is, folklore is exploded when it is carefully tested. While medical procedures are usually not simply folklore, they frequently do not work when they are used on new samples; in an ideal world, such careful investigation would precede the public availability of any therapeutic method and the Hippocratic adage, "First, do no harm," would be easier to practice. Such research has already caused critical reexamination of a number of "standard therapies" and, more important, has raised serious questions about the value of many medical tasks which are routinely performed and paid for as "scientific medicine."

The comparison of health statistics among wealthy, large-scale societies produces some puzzling, indeed challenging, results. Why should the life expectancy of Swedish men in the middle years of life be several years longer than that of males in England and the United States? Why should infant mortality in Sweden be near the biological minimum and much lower than in other high-energy societies? What are the true conditions for health and illness? Our growing familiarity with cross-societal comparisons does not answer questions, but it raises them in a dramatic fashion. Such broad differences indicate that the quality of medicine is not the sole, perhaps even the major, determinant of health as measured by these data (Anderson, 1972).

In truth, such analyses tend to place the role of the doctor within a much broader framework than he is used to. The quality of medical services and the uniformity of their delivery to a population is one parameter, within the more basic parameter of population health. On one hand we have surveys designed to screen populations for health problems; they usually find that one-fifth or more of the population studied has medically identifiable but untreated disease. (Thus a colleague defines a healthy person as "one who hasn't been studied enough.") On the other hand, a number of studies have indicated that physicians with their clinical, or pathological, orientation tend to follow the un-Hippocratic adage: When in doubt, practice medicine. A substantial amount of unnecessary treatment follows (McCarthy and Widmer, 1974; Wennberg and Gittelsohn, 1973).

Kerr White (1975) defines epidemiology literally as "that which is visited upon the people." In the case noted above, one might say that "the people are visited with an epidemic of surgeons"; nothing predicts the incidence of surgery as well as the presence of surgeons in the neighborhood. The broader "epidemiological" approach, then, takes into account at least in principle *all* those matters affecting health and illness: diet, heredity, sanitation, as well as the quality and accessibility of health care in its various forms. Indeed, the epidemiologist is perfectly willing to see modern medical practices as themselves possibly harmful to health, along with other direct and indirect effects of modern industrial technology.

From such a point of view, the average practicing physician has a very narrow definition of the nature of medicine. As White (1975) put it:

> If traditional medicine emphasizes the one-to-one relationship between patient and physician, certainly the central transaction in clinical medicine, there is now need to recognize the additional collective relationship between all physicians, the entire health care establishment, and the people. This "epidemiological shift" recognizes the interest of clinicians in numerator data, i.e., sick patients who seek care, and the interest of demographers and sociologists in denominator data, i.e., the structure and dynamics of populations, but it goes further, for epidemiology is concerned with both parts of the fraction. The epidemiologist is concerned with relating the sick who receive care to those who are sick but do not seek or receive care; he is indeed, concerned both with all those who are sick and all those who are at the risk of becoming sick. The larger societal viewpoint is at the heart of the epidemiologist's interest in rates and ratios, in samples, in the definition of populations, in the use of comparable terms, definitions and classification schemes, in the application of standardization procedures that permit comparisons over time and place, and in the estimation of bias through the recognition of observer error, observer variation, institutional selection, and of the ubiquitous impact of the placebo and Hawthorne effects on most clinical transactions. Of all this the contemporary graduate of an American medical school is largely ignorant. Indeed, it is not unfair to

> say that he is largely non-numerate and largely uncritical; his capacity to judge what medicine and medical science can and cannot contribute to human welfare in relationship to human needs is meager at best.

It is clear that the authority of the physician to do exactly as he thinks best with patients on a one-to-one basis whatever the cost and whatever the effects is under siege. Those who attack our present way of thinking about health, practicing medicine, and evaluating the results are not populist yahoos interested only in attacking the guild; indeed, most of them are also MDs. They are, however, people who for one reason or another have to consider medical care and health as applying to all of the population, in short, as *policy problems* for the total society and its government. Their critique of medicine as it is now practiced focuses upon the basic lack of accountability of the medical industry to the total society which supports it so handsomely.

MEDICINE, GOVERNMENT, AND TYPES OF AUTHORITY

Those who have contributed to the *social medicine* movement (the term seems somewhat more exact than "epidemiology"), whatever their professional training, have notably been persons vocationally concerned with the health of total populations. They are to be found in positions of responsibility among university faculties of Public Health, the Public Health Service, and other state and federal bureaucracies responsible for health in the United States, and officials of the National Health Service in the United Kingdom and other societies whose governments have accepted the burden of regulating the conditions for the health of the total population.

As physicians trained with the expectation of clinical practice, having learned the norms of solo, service-for-fee medicine, they are in the classical position of "the marginal man." Their very position in society, the rights and duties prescribed by their roles, force them to see medical services within a broader framework than the needs of individual patient and physician, while they know and value the basic commitment most doctors make to the health of "their" patients. As marginal men they understand the point of view of the public official who sees health as another function to be planned-for, provided, regulated; as physicians they view medicine as a calling, demanding a devotion to duty akin to that of the priest. Such thinkers are among the harshest critics of the medical professions as they are now organized, yet they tend also to be firm believers in the intrinsic value of the physician's work (see Morris, 1957; McLachlan and McKeown, 1971; Cochrane, 1972; White, 1975).

They know that, however responsible the individual physician or hospital administrator may be, the medical industry as a whole is irresponsible. Many persons ill or in hazard of illness have no recourse to the "armamentarium" of scientific medicine; for those with such recourse, many find at great expense that there is no cure in that doctor's bag. Yet the cost of medical care in the United States has soared, and continues to soar; today it approaches 9% of the gross national product—some 60% higher, proportionately, than in the United Kingdom where universal free medical care is the norm.

While commitment to basic research on the biomedical causes of disease is firm enough, the professors of social medicine are skeptical of the technological miracle which would eliminate the "human factor" from the practice of medicine. Indeed, one school argues that with the knowledge and techniques we have acquired for controlling infant mortality, contagious diseases, and much of the hazard in surgery, we have reached a plateau in medical innovation. The observer must judge for himself—how much progress in "understanding" a disease without effective prevention or cure is worth how many scarce resources in talent, labor and money? In any event, progress in curing the illness that remains—of which the "degenerative" diseases produced by the wear and tear of living predominate—has slowed considerably since the pharmacological breakthroughs of the 1940s and 1950s.

Thus from the viewpoint of social medicine, the major medical problem today is the application of what we know to populations who need help in prevention or cure of physical breakdowns. The study and reorganization of the delivery of health services is, then, central to their concerns. Within the broader parameters of nutrition, lifestyle, and public prophylaxis against disease, "the collective relationship between all physicians, the entire health care establishment, and all the people" becomes the focus for attention (White, 1975). Such an approach has a basic political strength, for it views health care as a cost in scarce resources competing with other needs of the society, but one which must nevertheless be made available to all the people who can make their demands felt in a mass democracy.

The authority base of the social medicine movement is, then, quite different from that of the practitioner of scientific medicine. The *units* for which one is responsible are different—total populations, as against individual patients. The *scientific framework* is different—the indicators are rates of illness and recovery, rather than the intricate history of disease in one case. The *standards of achievement* are perhaps most different of all: it is taken for granted that all patients eventually die, and the key question is one of over-all morbidity and mortality rates, given other demands on the resources of the society. Eventually the population-based medical establishment will have to face, formally or informally, a key question: What are the acceptable grounds for birth or death? These are questions appropriate to those who are socially responsible for the relationship between the medical establishment and all the people; they obviously violate the code of the individual practitioner, committed as he is to doing all he can for his patient's persistence as a biological entity, regardless of cost. The conflict between these two quite different types of authority is the root cause of the medical crisis of our time.

The uncontrolled proliferation of expensive, cost-insensitive medical care has priced medicine out of the market—insofar as it was ever in a market. With no functional substitutes for medical care available, the state must take responsibility for planning and regulating this sector of the economy. The demand for medical services, answered in part through Medicare and Medicaid and satisfied through the existing fee-for-practice system, is in principle impossible for any society to satisfy. So we come to the final major difference between exponents of social medicine and the conventional practitioner—the *client.* The latter's client is the individual who can afford his fees; the client of social medicine is the population as a whole, its desires expressed not through the market but through the state. In brief, the client is the state.

PSRO AND THE SOCIETAL DEFINITION OF HEALTH SERVICES

We may see the Professional Standards Review Organizations as basically constituted by the interests and intellectual perspectives held by concerned and often competing groups. The very meeting of these interests sets the boundaries of the PSRO in social space; they are in no sense external to the organizations. Internal to the concept of PSRO are the doctor's role as he sees it, the patient's role, and the state's role as defined by protagonists of the government. PSROs will constitute, then, arenas within which such interests will contend, with results that are by no means predetermined. To use a kindred metaphor, the PSRO will be an operating theater: in the dramatic sense, the very principle of health and the principle of healing are at issue, together with the questions of how they shall be defined, by whom determined, and by whom managed.

As PSROs were born of compromise between the perquisites of the health professionals and the public demands for quality assurance and cost control, so they may be viewed as created social arenas in which conflicts may be resolved. The nature of the conflicts derives directly from basic antinomies in the American political culture; these preceded the present situation in the health field, and are manifest in quite other aspects of our social life.

There is first the ancient conflict between the national society and the local community, the "locals," and the "Feds." The very existence of a PSRO will be defined by many as an intrusion of alien standards and ways into the everyday life of an older order. There is also the traditional conflict between the private interest and public interest, the provider's right to maximize his interests versus the public's right to "reasonable" levels of quality and cost.

There is, finally, the increasingly salient conflict between professional rights and lay judgments, the MDs versus the patients. If, as George Bernard Shaw said, any profession is a conspiracy against the public, medicine as an epitome of the "profession" may expect to draw fire in an exaggerated way, but one not unknown to other professions and assorted mysteries.

These conflicts run throughout large-scale society, and throughout the present-day United States. Wherever there is an effort to craft public programs to remedy ills or create new values for all the people the divisions will appear as fault lines across the terrain where plans are to be executed. Urban Renewal, Conservation, the War on Poverty, faced the same set of divisions, as constraints or components of achievement.

The basic right to health services for all the people has only recently become generally accepted. Its acceptance caused a radical shift in our view of the existing health establishment, with outcry and rancorous defense from health professionals who never bargained for such a responsibility. This new, emerging, *societal* definition of health services requires that they be adequate, accessible to all, and fair priced so that all can afford them. This is certainly not true of the present system, evolved out of fee-for-service medicine, private hospitals, government subsidy and the dole: which is precisely why that system is under attack.

Having accepted such a definition of the "right to health services," however, we are faced with many questions. How much health care, of what sort, is necessary and desirable; how much can the society afford? What is competent medical service—in its technical aspects and its subtler social aspects? (Is the cure versus care dichotomy valid: can the two be separated?) What is a just price? It is increasingly difficult to find such a price in large-scale society, where so many prices are administered by private or public corporations. Finally, who shall pay? What health services shall come out of private income, what out of the public income? (42% of the cost is now paid out of public income.)

These are just some of the questions raised by our acceptance of health care as a universal right in American society. They are not easily resolved in principle, let alone in action; their resolution may come about, in some degree, from evolving organizations trusted with the function of professional review.

TWO WAYS OF VIEWING AN AGENCY

There are two (polar) ways of viewing a newly created agency, mandated to bring about changes to which the legislature aspires. The first is inherently conservative, giving great weight to limitations in the existential situation, and constraints from the existing organization of the polity and the market. From this point of view, a national health program is essentially a *modifier* of existing practices in the provision and funding of health care, and the efforts to advance the quality of health care are meant to maintain existing practices at existing prices. Thus the Professional Standards Review Organization is analogous to the "governor" on an engine, which sets limits upon the quality and cost of work the engine can do.

From this perspective, most of the rule-organized actions of the PSRO could be predicted from a knowledge of the environmental constraints on the society and the structure of polity and market. After all, they are seen as limits within which the agency must act. (Further constraints are, of course, to be expected from the friction, loss, and slippage of bureaucratic organization, as well as the incompetence, venality, and malice of specific individuals.)

The provisions for Professional Review now written into law are generally of this type, inherently conservative. Thus it is assumed that the existential situation will continue to be one of continually expanding investment in health. It is also assumed that the polity as it acts upon health resources and their distribution will continue to be subordinate to the market, having recourse to no competing source of medical care. The result would be, then, a market effectively monopolized by practice-for-fee physicians with MD degrees from schools accredited and staffed by other MDs.

From such assumptions, it follows that the nature of health practices would be judged by MDs against their present habits of judgment. The cost of care would be judged against the prevailing price in the monopolized market. The judges would be in all cases fellow MDs who would constitute earnest that collegial loyalty would protect the individual practitioner from "government regulation." At the same time, this self-regulation by fellow monopolists would prevent regulation by officials who did not share the same position, economically or intellectually. The organization of PSRO jurisdictions by existing political jurisdictions, e.g., states and counties, would make it feasible to use existing private organizations (medical societies) as the scaffolding for the Review programs.

An alternative way of seeing such an agency, however, is to define it as an "active site," from which basic changes in structure may emerge. By accumulating independent power, the agency may modify the constraints within which it was developed. Professional Review might effect changes in the norms of medical practice through improving communication among and about practitioners at work. Such change would require the translation of laws into rules, of rules into norms, through evaluation of practices and the publicizing of results. Instead of accepting the statistical mode as the desirable norm, such Review might be focused upon improving the outcome of health care through *shifting* the mode. Eventually such an agency might change the method of evaluating practice, the quality of health services, and the ways of pricing them.

Robert Alford (1973), in his analysis of the politics of health services (which he characterizes as "dynamics without change") sees three major protagonists in contention. First there are the professional monopolists, who try to protect a winning game and improve it if they can; second, there are bureaucratic rationalizers, who wish to make predictable the provision of health and integrate it into an orderly system; third, there are the organized consumers, who try to generate pressure on the system to improve accessibility, quality, and the terms of trade for themselves. From this point of view, the conservative definition of PSRO is a logical compromise between the interests of professional monopolizers and bureaucratic rationalizers.

Alford tends to dismiss the organized consumers as an important political force, for he sees them as lacking resources, continuity, and legitimacy. They carry on a rather ineffectual guerrilla war, while the major battles are still between the monopolists and organizers. And, viewing only the voluntary organizations of consumers interested in health care, this seems an accurate assessment. However, if we see consumers as increasingly organized through the mainline political process, with voice in Congress, membership on major boards, and representatives in the governmental bureaucracy, their interests no longer appear so negligible. If, further, we see this political representation resting upon an increasing federal interest as citizens pay the health bill and receive the health care through the government, consumers so organized may become the dominant force.

In PSROs then, the consumers are represented as both taxpayers into the government's medical program and as potential patients. The rationalizers are intent on creating an orderly system in which accountability to the government will be guaranteed while the professional monopolists try to protect their practice from laymen. We can specify some possible outcomes, looking at the goals of each party, rationalizers and monopolists, as variables ranging from a positive to a negative pole with respect to the interests of the consumers.

The aims of the rationalizers may be satisfied through the creation of PSROs which are genuine "active sites," arenas devoted to improving standards of health care delivery and standards of pricing. They may also be satisfied through developing structures which are simple governors, limiting and determining the standard of practice for the standard price. Finally, they may be nothing more than monitoring and recording devices, which regularize the status quo until the next collective bargaining session. This extreme pole would be Richard Snyder's "negative model."

The aims of the professional monopolists are, at the positive pole, to guarantee their right to practice "absolute medicine," to prescribe and/or execute for each patient exactly what is called for in their judgment. On the negative side, the result would be the practice of routine medicine, of defensive medicine, even of minimal medicine. This is the "negative model" for the public of what can be expected from this interest group. These notions may be schematized as follows:

TABLE 1
INTERESTS AND OUTCOMES

	Bureaucratic Rationalizers	
Professional Monopolists	**"active site"**	**"governor"**
"absolute medicine"	A	B
"minimal medicine"	C	D

Should the rationalizers and the monopolists succeed in achieving alternative "A," it would be the best of possible outcomes for all three parties (keeping in mind the government as agent for organized consumers). The best possible medicine would be practiced within a framework which satisfied the needs of rationalizers for cost accountability and of consumers for quality assurance. Furthermore, the PSRO would be a dynamic which, from a continuing evaluation of medical services, laid the groundwork for improvement.

Alternative "B," the negative model of rationalizing, would have no effect on the professional monopolists, while in the postulated "C" medical care would worsen, for the very activity of the PSRO would encourage routine and defensive medicine. Alternative "D," however, would be the worst possible outcome. PSROs would substitute conventionalized record-keeping for genuine monitoring of health care, creating symmetrical systems of statistical lies which satisfied both the professional monopolists and the Office of Management and Budget. The monopolists, in turn, could trim their behavior to the minimum compliance required by the statistical system, (punishing a few scapecoats to satisfy numerical requirements) and exercising complete freedom outside the areas tapped by the reporting forms.

This is the grimmest vision of the total enterprise. The professional's responsibility would be transformed into the sheer power of status, underwritten by PSRO; free of concern with duties, he could concentrate upon his rights. That most valuable and powerful of social resources, trust, would be further eroded. The growing distrust of providers by consumers, of the state by the providers (and vice versa) would increase steadily, for Professional Review would provide little reason for confidence in either intent or execution.

As matters stand there is little evidence that the societal definition of acceptable health services is being fulfilled. Distrust of the competence and benevolence of health providers continues to grow and, with increasing investment in health by the government, the need for accountability is manifest. Let us consider the possibility of achieving "A," the PSRO as an active site encouraging the practice of absolute medicine within a framework of social responsibility.

PSRO AS AN AGENT OF CHANGE

Given the high probability that health professionals will continue to be well rewarded, that medicine will continue to be at best a very inexact practice, the question is: How can we change the terms of trade? How can public and *patient* be sure incompetence will be reduced and pricing reflect what needs to be done at affordable costs? Nothing less will restore the trust that is needed between healers, patients, and the public.

PSRO as an agency of change establishes, for the first time, *standards of practice* which are socially validated and supported by the authority of the state, administered through responsible physicians. Standards for health care, even at the mundane level of length of hospital stay required for good medical practice, are of major importance as precedents; they allow accountability. Standards for health prices are equally important, and the relation between prices and adequate practice is absolutely necessary for any equitable system of exchange. Whether one calls such standards the establishment of the principle of accountability, or the entering wedge of government control, it is clear that they represent a major innovation in American health services.

The statutes require the creation of the agencies, and therefore they require certain conforming behavior by the physicians and other health professionals. This may be called the *situational* determination of change. Behavior is different. Again, at the mundane level, hospital discharges are handled in a more orderly fashion, admissions are more carefully considered. The argument against the value of such change is simple: that it does not really affect basic belief. One may say it changes the periphery and leaves the core values untouched and, as long as this is the case, health professionals will use the new constraints to protect and extend the freedoms they deem good, necessary, and right for their professional role.

Donald A. Schon (1971) has termed the problem one of "dynamic conservatism." By this he means that when change agents attack a problem, the components of the problem actively reorganize to resist or absorb change so that the basic situation remains as it was. The effort to depart from the steady state is frustrated by the objective power, flexibility, and determination of the actors who value that state of affairs. So the key question in determining the effectiveness of PSROs is whether situational requirements, stemming from the authority of the state, will stimulate a departure from the steady state of medicine as it is practiced today.

One possibility is that, through interaction, the core values will legitimate the situational constraints, and the constraints make clear and objective the nature of core values. Thus the process of review may place PSROs in the position of highlighting unneeded treatment, as in tonsillectomy for children, and *allow* the surgeon to say "no" with finality to the parents. Or to the contrary, commitment to the values of "absolute medicine" may greatly extend, strengthen, and differentiate the criteria which are used in making evaluations of medical care.

In discussing the possibilities for change of a truly basic nature, Schon (1971) specifies several requirements. There must be a new product which has comparatively greater value than the old; there must be an expanding market for the new product in which this advantage tells; there must be an organizational base from which the new product is dispersed, sold, and supported. His examples are most telling in the cases of commodities such as steel, but would seem to apply to health services. In the latter case, we may speak of socially distributed, accountable health services, compared to privatized service-for-fee, as *product*; the electorate as patients is a rapidly expanding *market*; the federal government as regulator of health services for the population as a whole is the overall *organizational base.*

But we have seen that we do not want to displace one product by another and quite different one. Alternative "A" combines active social responsibility of the PSRO with the technological responsibility of the physicians; we wish to integrate the two value systems. In that process, the PSRO model may have unexpected advantages.

I have noted three tensions which run through American society: they may be indicated by the short-hand "local-national," "private-public," and "professional-layman." In any large enterprise each type of dichotomy is apt to be present, for it is a large scale but highly interdependent social system; still, the relative power and freedom of the paired terms will vary by the case and over time. This is not a judgment of "pathology"; such antinomies are to be expected. The consequences of given balances may, of course, be quite destructive.

The combination of national rules and local actors, national statutes and local interpretations, may be conducive to the integration of value systems. While there is ground for distortion in local control, there is also an opportunity to invoke genuine support for both the standards and the standard making and enforcing processes, *if* they are locally developed and staffed. If adapting standards to local physicians' preferences helps to bring about serious applications backed by the moral judgment of the professional fraternity, the price may well be worth it. Given the amount of uncertainty in many areas of medical knowledge, diversity may even be an advantage to the larger enterprise of medical practice, particularly if it is publicly, and accurately, documented.

As for the division between private and public interest, let us remember that it is endemic to society. In the PSROs the doctor's dilemma, how to price vital services when there is neither market nor bureaucratic rule, may be resolved in a way he can accept. After all, the interests of the state in cost accountability will be applied impersonally across the board to all work; this allows individuals to reduce their own uncertainty without paying a moral cost. In the same way, hospital personnel can be reinforced in their own professional values, which certainly do not include providing unnecessary and extravagantly priced room and board to unwilling guests. While abuse is always a possibility (the risk is the price of delegation), we must remember that there is today no publicly known and accepted "just price" in health services. PSRO may be a device helping to create such a price system.

With respect to the tension between the professional and lay perspective and interest, PSRO could perhaps be regarded as a relative gain in any event. However, insofar as the use of profiles is sensitive and medical care evaluation is well-focused and executed, the professional basis for judgments should be strengthened in a publicly demonstrable fashion. As medical care evaluations improve in technique, using methods developed in random clinical trial, our knowledge of efficacious and cost-effective practice should accumulate. Whether this lowers the cost of medical care or not, it should lower the waste motion and money, which is a principal moral concern of all involved. The obverse of waste is, of course, increasing value in the health care that is given; better treatment for patients.

The Professional Standards Review Organization, whatever the intent of the Congress, is apt to be an organization which expands. It should be remembered that there are at least three levels of review, all of them *together* addressing the requirements for social medicine: that it be adequate, accessible, fairly priced, and paid for. These levels are (1) review of specific medical care and providers; (2) review of the standards themselves; (3) review of the PSRO operation as a monitor and guardian of both practices and standards.

The review of specific practitioners and organizations against given standards rests upon the moral norms of the professional community and its pride; this is why the standards must make sense to the participants. At the same time, the standards must themselves be reviewed against more general criteria, including the range of practices in various PSRO regions. Most importantly, they must be evaluated against the outcomes of given practices. Whether this is done through PSRO or other agencies, the scientific solution to the problem of standards obviously requires randomized clinical trial. Such a procedure is a conventional resolution for the health professional, for it rests upon scientific test and evaluation. At the same time, it satisfies the anxieties of the patient, who seeks health care only because he believes in its efficaciousness; it satisfies the interests of the state for it is the best earnest of moral commitment to high quality medicine.

PSROs must also be subject to review, for they are liable to all the pathologies of bureaucracy and some all their own. The social perpetual-motion machine has not yet been invented; one which is frankly meant to monitor and control behavior, apply standards and assess sanctions, within a context of differing opinions and interests, is certainly going to need careful scrutiny. Such review of the reviewers rests upon the norms of social science; propositions about what is the case in society, what "works" to produce what, are subject to the rules of publicity and reversibility. The means by which the proposition is judged must be made public, and all propositions are subject to reversibility. PSROs, like much of human life, are extraordinarily complex hypotheses.

REFERENCES

ALFORD, R. (1973). The political economy of health care. Warner Module 96.

ANDERSON, O. W. (1972). Can there be equity? The United States, Sweden, and England. New York: Wiley-Interscience.

COCHRANE, A. L. (1972). Effectiveness and efficiency. London: Nuffield Provincial Hospital Trust.

COTTRELL, F. (1955). Energy and society. New York: McGraw-Hill.

DUBOS, R. J. (1959). The mirage of health. New York: Harpers.

FRIEDSON, E., and LORBER, J. (1972). Medical men and their work. Chicago: Aldine, Atherton.

GREER, S. (1962). The emerging city, myth and reality. New York: Free Press.

LANGER, S. (1957). Philosophy in a new key. Cambridge, Mass.: Harvard University Press.

McCARTHY, E. G., and WIDMER, G. W. (1974). "Screening by consultants on recommended elective surgical procedures." New England Journal of Medicine (December 19): 1331-1335.

McKEOWN, T., and LOWE, C. R. Introduction to social medicine. Oxford: Blackwell. (1966).

McLACHLAN, G., and McKEOWN, T. [eds.] (1971). Medical history and medical care. London: Oxford University Press.

McNEIL, W. H. (1970). The rise of the west. Chicago: University of Chicago Press.

MORRIS, J. N. (1957). The uses of epidemiology. Edinburgh: Livingston.

SCHON, D. A. (1971). Beyond the stable state. New York: Norton.

SNYDER, R. C. (1975). "Peer review and health care: An exercise in policy clarification." In Proceedings, Conference on Professional Self Regulation, S. Greer (ed.). Washington, D.C.: H.E.W., June.

SIGERIST, H. E., (1975). Civilization and Disease. College Park, MD.: McGrath.

WENNBERG, J., and GITTELSOHN, A. (1973). "Small area variations in health care delivery." *Sciences* (December 14): 1102-1108.

WHITE, K. L. (1975). "Opportunities and needs for epidemiology and health statistics in the United States." Presented at the Invitational Conference on Epidemiology, Baltimore, Maryland, March.

Part II

Accountability and Political Institutions

Introduction

□ IN A DEMOCRACY THE PROBLEM of political accountability centers on the evaluation and control of the outputs of the institutions granted access to the authority of the state. The chapters in this section all focus on the accountability of public institutions. Each of the authors addresses different aspects of accountability, yet it is obvious that several impediments to accountability are common to all political institutions. Indeed, one is struck by the minimal differences which exist among nominally dissimilar institutions.

In "A Reconceptualization of Legislative Accountability," Hedlund and Hamm address several major issues in the accountability of legislative institutions. Understanding accountability to mean an assessment of legislative goals and movement toward those goals, they develop a variety of legislative performance indicators. The authors stipulate seven major operative goals and functions for the legislative process and proceed to outline operational approaches for empirical research. Hedlund and Hamm conclude with suggestions for increasing legislative accountability, suggestions which would increase the ability to judge legislative performance through the collection and publication of performance indicators.

David Rosenbloom, in a far-ranging critique of the administrative state, identifies several impediments to bureaucratic accountability in the next chapter. Because of factors such as personnel practices, constitutional rights of public employees, public sector labor unions, and the like, control over the bureaucracy is virtually impossible. Rosenbloom suggests that the primary means for achieving bureaucratic accountability, if indeed it is desirable, may be found in making the bureau a more respresentative institution.

The next two chapters provide a picture of judicial institutions which is analogous to that of the bureaucracy. It is argued that the courts are significant, yet uncontrolled, policy-making institutions. Gibson has attempted to measure the policy outputs of trial courts in a fashion not unlike that proposed by Hedlund and Hamm. He demonstrates that the policies made by trial courts have significant implications for the authoritative allocation of values in society, and that these policies mirror, to some extent, the political and social complexion of the jurisdictions of the courts. That the relationships are not stronger, it is argued, is to some degree a reflection of the view that courts are apolitical institutions.

Wasby explores this point in much greater detail in chapter six. Rejecting the notion that courts are totally unaccountable, he argues that courts are at least partially accountable to both the legal system and the political system through several different processes or mechanisms. While accountability to the legal system may conflict with accountability to the political system, Wasby concludes that there is little reason to believe that courts are any more unaccountable than any other political institutions.

Finally, Virginia Gray offers a perspective on the accountability issue which is at odds with the first four chapters of this section. She argues that because policy frequently transcends the boundaries of single institutions, the policy *process* is the appropriate focus of accountability. Courts, legislatures, and bureaucracies are all concerned with the control of crime, for instance, and the institutional approach is likely to produce a segmentalized view of the accountability of government policy. She further argues that different types of policy require different conceptualizations of accountability. Her chapter demonstrates the importance of the conceptual definition of accountability for any evaluation of public policy.

3

Reconceptualizing Legislative Accountability

RONALD D. HEDLUND
KEITH E. HAMM

☐ ONE OF THE MORE PERVASIVE political concerns among citizens, academics, and public officials is the issue of public agency accountability in an increasingly impersonal, urban society. The turbulent and troubled political scene of the last decade, with its large-scale, anti-government demonstrations, political scandals, and an apparent inability of government to deal with important issues, has fostered heightened public cynicism and distrust. As a result, individual citizens and public-minded groups are calling for broad personnel, structural, process, and political changes to increase the accountability of all types of public agencies and public servants to the people being served.

Although this movement toward greater accountability strikes a responsive chord among many, serious problems confront any meaningful discussion of the topic. Nagging questions continually arise:

(1) What is accountability? (Is accountability achieved by having certain types of *formal structural arrangements* that limit a public

AUTHORS' NOTE: *The authors wish to acknowledge the assistance of several persons in completing this paper. Principal are John G. Grumm, whose proposals for measuring policy output form the base for the indicators suggested here, and E. Terrance Jones for his ideas regarding the media's role in disseminating information. Financial support has been provided by the University of Wisconsin-Milwaukee, Graduate School, and the Urban Research Center. Special thanks go to our wives.*

official's power, require periodic review of his/her performance, or provide for removal from office? Or, does accountability require that attention be given to *individual behavior* and *organizational actions* as these relate to the spirit and tenets of responsibility to the public?)

(2) Who should be held accountable? (In a complex, bureaucratized, and impersonal governmental system, is the *individual office holder* or *agency person* the singular target for accountability? Or, given that most public officials are only one small component in the decision-making process, should the *entire governmental organization* or *apparatus* be the target of accountability strategies?)

(3) To whom should accountability be directed? (Is accountability to be directed to *all citizens,* whether or not they are all affected by the actions or policies? Or, should accountability be directed toward smaller more directly concerned segments? If so, to which segments should accountability be directed—to a public official's own *constituency,* to those persons *directly affected* by an action or decision, or to the *media* whose power of focusing attention or highlighting an issue gives them substantial informal powers for directing accountability?)

(4) Accountability for what? (Are there any limitations on the behavior or actions which should be the concern for accountability? Should accountability be directed only toward the *policy decisions* that are made, i.e., the end result of the decision process? Is a public official's entire *personal behavior* appropriate for the purpose of accountability? Or, are the *overall actions of a governmental body* the proper subject for accountability reviews?)

(5) How much accountability is desirable? (Should public officials and organizations be held accountable after *every action,* a kind of ever-present, ongoing plebiscite? Or should accountability reviews be held *periodically,* such as annually, every two years, or at five-year intervals?)

Unfortunately, little agreement appears in current discussions regarding these aspects of accountability. For example, advocates of increased public accountability seem to agree only

> that it is important to have rather full disclosure of the choices made by public officials and that it is important to have provisions for their removal from office. Beyond this, the advocates of public accountability part company. [Francis, 1972:184]

It is toward resolving some of these questions as they apply to legislative organizations that this chapter is addressed. Given the unique organizational nature and decision-making process used in legislative bodies, this chapter will begin with a discussion of accountability as it relates to the basic democratic tenet of representation. One must initiate the discussion here because concern with accountability in legislative bodies had its origin in competing notions of representation. We shall then consider alternative mechanisms which have been used to create and insure accountability in legislative bodies. A proposal for viewing accountability in terms of institutional performance using certain pertinent measures will then be presented. And finally, our discussion will focus on alternative strategies which may be pursued to increase levels of accountability.

ACCOUNTABILITY AND REPRESENTATIVE GOVERNMENT

Representation—one person or group acting in an official capacity on behalf of some others—is a basic principle in our scheme of government. Whenever one considers political representation, one is confronted with the twin elements of authorization and accountability. Initially, the concern is with the formal arrangements surrounding the authorization of the representative.

> a representative is someone who has been authorized to act. This means that he has been given a right to act which he did not have before, while the represented has become responsible for the consequences of that action as if he had done it himself. It is a view strongly skewed in favor of the representative. His rights have been enlarged and his responsibilities have been (if anything) decreased. The represented, in contrast, has acquired new responsibilities and (if anything) given up some of his rights. [Pitkin, 1967:38-39]

Authorization involves transferring the power to make decisions on behalf of a person from that person to someone designated a representative. The represented gives to the representative the right and authority to make decisions, enter into agreements, and generally act on behalf of the represented.

In a political sense, authorization involves a *grant* of power from the represented to the representative and stems from the realization that direct involvement in governmental decision making is not always feasible in a large-scale, mass-based, impersonal social system. Whenever a person or organization is empowered to *act on behalf of other citizens,* representation in terms of authorization is evident. Decisions made in this fashion become binding for all citizens, whether an individual agrees with the decision or not.

> The legal essence of representation is that the representatives—whatever the manner of their investiture—are authorized in advance to act conjointly on behalf of their constituents and bind them by their collective decisions. [Loewenstein, 1957:38]

Contrasted with authorization in various discussions of representation is accountability. Here, the emphasis is on the representative answering for his actions. In this sense, accountability entails special obligations. It is a post hoc means for assuring that the representative knows that sooner or later, in some way, the represented will pass judgment on the representative's actions and will be able to remove the representative from this position. Accountability requires a public official or political institution to

> periodically face those to whom it is responsible for a judgment upon its record. If the decision makers are found wanting, the "rascals" can be "turned out," removed from office . . . legislators must calculate the popular response to their actions (or lack thereof) or risk loss of position and power. [Rieselbach, 1975:69)

Maintaining a balance between authorization and accountability is a continuous process in any representative democratic system.

ACCOUNTABILITY LINKAGES

Five primary linkages have been posited as means to achieve accountability—constitutional restraints, internalization of norms, electoral linkages, open process/structural change, and performance review. We shall consider each briefly.

CONSTITUTIONAL RESTRAINTS

One method for advancing accountability is through constitutional means, as checks and balances and separation of powers. This involves implementing formal countervailing powers which permit officials in one branch of government to control possible abuses by individuals who occupy official positions elsewhere. Structural devices which limit the power held by any one branch or level of government do this. Further restraint across branches and levels is expected to occur because various kinds of people are represented "by different branches of government and differing levels of officialdom" (Eulau and Prewitt, 1973:445). In actuality, this method of accountability involves political officials checking on other political officials. However, there is no guarantee that various kinds of citizens are being served by different levels and branches, at least not as envisioned by the architects of this system. Further, the expansion of informal ties across branches and levels of government reduce the likelihood that any one branch can effectively check another, at least in terms of strict accountability (Eulau and Prewitt, 1973:445; Ippolito et al., 1976:133).

INTERNAL ACCOUNTABILITY

The basic thesis of internal accountability is that

> political elites (public officials as well as activists) will behave in a responsive and responsible fashion because they have internalized these values. Ideally, *political elites hold themselves accountable* and do not require external checks. [Ippolito et al., 1976:134]

In this approach certain attitudes, such as responsiveness, are to be internalized and practiced by public officials as they exercise power. The implication is that

> the elected leaders are solicitous of the views of the masses not because the elected fear voter sanctions; rather the governors remain sensitive to shifts in public sentiment because that is what men holding public office *are supposed to do.* [Prewitt, 1970:123]

Recent evidence indicates that uniformity of such attitudes and behavior among political elites is not that pervasive; in fact, after the electoral restraints have been removed, many public officials behave in a manner considered contradictory to democratic norms of responsiveness and responsibility. A greater difficulty with this thesis, however, is that it

> suggests that the criteria by which governors are to be evaluated, and are in fact evaluated, are set by the governors themselves or at least by their predecessors. [Eulau and Prewitt, 1973:446]

The result is that the question "accountable to what" is answered as accountable to the values which are mainly shaped and molded by those individuals who are in a role to determine the political norms of society (Eulau and Prewitt, 1973:446).

ELECTORAL LINKAGE VIA THE INDIVIDUAL REPRESENTATIVE

Almost every discussion regarding the accountability of elected public officials centers around the electoral process. As pointed out by Kuklinski in chapter nine, when standing for re-election, the individual office holder presents his or her performance record for review and thus is held accountable. Any elected public official readily admits that s/he is constrained in his/her actions by the need for reelection at some future date. However, as Kuklinski indicates, the electoral linkage also poses difficulties for establishing accountability.

OPEN PROCESS AND STRUCTURAL CHANGE

A somewhat more recent approach to establishing accountability, taken by certain public-spirited, reform-minded organizations—e.g., Common Cause and Legis 50 (formerly the Citizens Conference on State Legislatures)—argues that if the legislative process is open to public and press scrutiny and if the organization operates in certain ways, then accountability and favorable citizen evaluation will result.

> If the press and public are to improve their understanding of what the legislature is doing, and thus to hold it accountable for what it does, the legislative process itself must be understandable and, at every important step, open to public view. Understanding is the first step toward accountability. [Burns, 1971:79-80]

Unlike accountability via elections, this approach provides a means for establishing accountability for the *entire legislature* and not just individual legislators. Again, though, certain critical shortcomings exist. Implicit is the assumption that if you somehow open up the legislative process and make it understandable, you will improve accountability and citizen evaluations, but the precise tie between openness and accountability is vague.

More damaging, however, is some fragmentary evidence regarding the relationship between measures of openness in the legislative process and public evaluations. Based on this open process/structural change perspective, one would expect that those states having a high degree of accountability, openness, and understandability would also be those states having relatively higher levels of citizen evaluations represented by their satisfaction with the specific activities of the legislature. Table 1 analyzes the states for which data are available, citizen judgments on specific legislative performance (and state government actions), and the Legis 50 ranking on accountability. Summarizing Table 1, there appears to be a negative or insignificant relationship between measures of accountability and performance. For example, those state legislatures having higher levels of accountability actually have a lower percentage approving the legislature's performance (rho = –.60), a higher percentage who feel that the important decisions made by the state government are made by a few men who don't hold public office (rho = -.21), a greater percentage who have least confidence in state government relative to the national and local governments (rho = -.61), and, finally, no relationship with the percentage of people who feel that state government pays a lot of attention to what the people think when it decides what to do (rho = -.11). In other words, those legislatures adjudged to be the most accountable in terms of openness, understandability, and influence capability actually have lower citizens' favorable predispositions toward them. This conclusion is tentative, of course, given the single point in time evaluation. But it seems that, if anything, exposing the inner workings of the legislative process to the citizenry, granting access, and so on, *without also improving the functioning of the legislature only leads to greater feelings of unresponsiveness.* The potential impact of such a strategy is very evident in two evaluations of problems with state legislatures. In Minnesota in 1965, over one-third of the respondents in one study thought that the legislature wasted time; while in Iowa during 1968, 59% indicated that legislators wasted time, often focusing on the trivial at the expense of more important tasks (Patterson, 1976:163). The public, in these studies, was concerned with performance–the speed, efficiency, and volume of legislation disposed of by the legislature rather than openness and access. It seems, then, that a more viable approach to accountability would emphasize a legislature's ability to accomplish certain tasks and to pursue agreed-upon goals *without* wasting time and other resources.

PERFORMANCE REVIEW OF THE INSTITUTION

The fifth approach, and the one which we will use here, views institutional accountability as understandable in terms of some overall *assessment of legislative goals* and *movement toward achieving these goals.* In the terminology of organizational research, this approach involves a focus on organizational performance, and herein "accountability means that there are standards against which the performance of the officeholders can be measured" (Eulau and Prewitt, 1973:444). As with the preceding approach, performance review concentrates on activity at the *organizational level* thereby concentrating on establishing *institutional accountability;* however, individual level accountability is an aspect of this approach as well.

At present, however, there are no widely accepted approaches or indicators for evaluating legislative performance. As Fenno (1975:279) indicates in his discussion of the differing criteria for evaluating individual representatives and the institution,

> The individual legislator knows when he has met our standards of representativeness; he is re-elected. But no such definitive measure of legislative success exists.

TABLE 1
STATE RANKINGS ON LEGISLATIVE ACCOUNTABILITY AND ATTITUDES ON LEGISLATIVE CONTROL AND CONFIDENCE

	(1) Accountability Rank[a]	(2) Approve Job Done by Legislature[b]		(3) State Decisions Made by Few Non-officials[c]		(4) State Government Pays Attention to What People Think[d]		(5) Most Confidence in State Government[e]		(6) Least Confidence in State Government[f]	
		Percent	*Rank*	*Percent*	*Rank*	*Percent*	*Rank*	*Percent*	*Rank*	*Percent*	*Rank*
Massachusetts	9	32	13	29	7	24	11.5	7	13	36	13
New York	6	29	14	23	3.5	24	11.5	10	12	14	6
Pennsylvania	7	40	10	31	10	23	13	13	11	19	9
Illinois	2	44	9	30	8.5	30	8.5	14	10	20	10
Minnesota	3	49	4.5	19	1	43	3	19	6.5	15	7
Ohio	8	55	2	25	5.5	41	4.5	20	5	17	8
South Dakota	5	49	4.5	23	3.5	50	1	26	3	11	3.5
California	1	33	12	36	12	31	7	17	8	22	11
Florida	4	37	11	37	13	27	10	16	9	23	12
North Carolina	11	51	3	25	5.5	33	6	29	2	7	1
Texas	10	46	8	33	11	41	4.5	19	6.5	33	5
Alabama	13	56	1	20	2	45	2	41	1	8	2
Louisiana	12	47	7	30	8.5	30	8.5	25	4	11	3.5
Spearman's rho, (test for difference of ranks)		Col 1 and 2 rho = –.60		Col. 1 and 3 rho = –.21		Col. 1 and 4 rho = –.11		Col. 1 and 5 rho = –.46		Col. 1 and 6 rho = –.61	

And, the criteria that do exist for judging institutional performance appear to be applied inconsistently (Fenno, 1975:279). What is needed, then, is some valid, empirical, objective criteria consistently applied in order to evaluate legislative performance.

The approach developed in this chapter involves identifying various agreed-upon goals or activities being pursued in a legislature. Once these goals have been identified, this approach requires extensive specification of the dimensions and indicators for each goal. Then data can be collected for each indicator at several differing points in time so that an assessment can be made regarding movement toward or away from goals. This paper will concentrate on identifying these goals and suggesting appropriate indicators; collecting pertinent data will be pursued elsewhere (see Hedlund and Hamm, 1977, 1978).

LEGISLATIVE PERFORMANCE AND ACCOUNTABILITY

Inherent in all discussions of governmental accountability is a concern regarding the range of behaviors and activity for which representatives and organizations will be held accountable. Most philosophical and legal treatments use the word "actions" as the definition of range. Sometimes this has been translated into meaning accountability only for policy decisions. Public officials, as well as political organizations like legislatures, perform many functions in addition to making public policy. The position taken here is that accountability must be broadly understood and must be measured in terms of a wide range of *actions*, not just the final decisions regarding *which policy alternative* is implemented.

A review of classroom texts and other materials on the American legislative process produces several somewhat different lists of actions which con-

---

[a]Rank order on accountability (Citizens Conference on State Legislatures, 1971:40).

[b]Percent of responses, "Excellent" and "Pretty Good," to the question "In general, how would you rate the job the state legislature has done in the past two years?" (Black et al., 1974:186).

[c]Percent of responses, "Agree," to the question, "Most of the important decisions made by the state government at (name of state capital) are *actually* made by a few men who don't even hold a public office" (Black et al., 1974:193).

[d]Percent of responses, "Agree," to the question: "The state government at (name of state capital) pays a lot of attention to what the people think when it decides what to do" (Black et al., 1974:190).

[e]Percent of responses, "State," to the question, "Do you have *most* faith and confidence in the *national* government, your *state's* government, or in *local* government around here?" (Black et al., 1974:188).

[f]Percent of responses, "State," to the question, "Which level of government do you have the *least* faith and confidence in—the *national* government, your *state's* government, or *local* government around here?" (Black et al., 1974:188).

stitute legislative goals and functions (Keefe and Ogul, 1977; Jewell and Patterson, 1977; Van der Slik, 1977; Saloma, 1969; Blair, 1967). Comparisons across such lists, however, do suggest agreement on seven goals or functions—policy responsiveness; formal decision-making; oversight of the administrative branch of government; constituent representation in policy formation; constituent relations, education and advocacy; solidarity building; and problem investigation. This list can be seen as dimensions for evaluating legislative performance. Some dimensions are derived from institution-based data, while others involve individual level information. Each is important for measuring accountability. Accountability then involves assessment to determine the legislature's success in achieving these. (This approach is based upon that suggested by Katz and Kahn, 1966; and Price, 1968.) An elaboration of each of these goals or functions follows.

POLICY RESPONSIVENESS

Legislatures are judged according to the specific output of the enactment process—the nature of the policy decisions. Because the allocation of values and resources resulting from policy decisions is uneven, any decision is likely to benefit some sector of society at the expense of some other. Thus, a legislature must face the prospect that some groups of citizens will be dissatisfied with the decisions being made while others will be pleased. Nevertheless, in making decisions, legislators frequently are responding to the needs and demands made upon them.

> To be responsive, a decision making center like a legislature must listen to, and take account of, the ideas and sentiments of those who will be affected by its policy decisions or pronouncements. [Rieselbach, 1975:69]

FORMAL DECISION MAKING

In the policy-making process, legislatures make decisions on public policy matters, and these decisions typically take the form of laws being enacted or refused, money being appropriated, and resolutions being adopted. Decision making in these forms rests upon some grant of power in a constitution or charter. Such a basis for this power gives legislatures the legitimacy necessary for their role in public decision making. Admittedly, legislators may not initiate the ideas for all or even much legislation and may not even be responsible for drafting the ideas into bill form; however, legislative action is critical because legislatures are authorized by constitutions *to make binding decisions* for society. Specifically, legislatures review proposals for action and modify them according to the legislators' evaluations. Legislatures also vote final approval on all laws, giving them legitimacy to become official public policy. Any list of legislative goals or functions must include this decision-making role.

OVERSIGHT OF THE ADMINISTRATION

Embodied in our system of government on both the state and national levels is the principle of divided power. Through the separation of power among the three branches of government and the elaborate system of checks and balances, each branch and agency is provided with some but not exclusive political power. One important aspect of this system is legislative oversight of the executive branch of government. Legislative oversight refers to the legislature's responsibility for reviewing actions taken by executive agencies and for initiating proposals to counteract or confirm their decisions.

> the legislature has a long-established concern with inquiring into administrative conduct and the exercise of administrative discretion under the acts of the legislature, as well as with ascertaining administrative compliance with legislative intent. In the usual phrasing, the legislature's supervisory role consists of questioning, reviewing and assessing, modifying, and rejecting policies of the administration. [Keefe and Ogul, 1977:20]

As the scope of governmental activity increases, this oversight function seems to be receiving increased attention, also.

CONSTITUENT REPRESENTATION IN POLICY FORMATION

Inherent in any representational system is the degree to which the represented's views are embodied in the represented's decisions. Because of the geographical basis for selecting representatives, this function frequently involves advancing the district's views in decision making. Such involvement can vary from largely symbolic gestures as one saying he "represents the views of his constituents" to very specific acts intended to advance the pecuniary interests of constituents.

CONSTITUENT RELATIONS, EDUCATION, AND ADVOCACY

In addition to reflecting constituency interests in policy formation, our political system has developed the expectation that legislators will serve as personal envoys of and advocates for their constituents. Commonly referred to as the "errand boy" function, this activity involves varying efforts by representatives to intercede in differing ways with agencies on behalf of constituents. At one extreme, this may involve a simple request for providing information to a constituent; while at the other, it may require extensive personal effort to untangle a complex problem.

Various legislative efforts may also be directed toward educating constituents and the general public. Through legislative hearings, official reports, public speeches, and floor debate, the legislature can raise the salience level of an issue and provide added information for public edification.

SOLIDARITY BUILDING

Legislatures also provide a means whereby citizens can approach government with ideas, policy preferences, and opinions. Through contacts with individual legislators, appearances at committee hearings, and involvement in group activities, individuals are able to articulate their views to legislators. As they are involved in such activities, citizens may begin to feel that they are an integral part of the political system and that they have some opportunity to influence the course of public policy. If the availability of this means for approaching government reduces tensions regarding the conduct of policy formation and fosters regard for the political system, legislatures may be performing a type of safety-valve function. By reducing divisions among people and generating positive attitudes for the political system, the legislature is building solidarity.

> Any political system is subject to stress. Groups that are dissatisfied with the way in which conflicts have been resolved may become alienated from the system and unwilling to accept the political decisions emanating from it . . . The legislative system functions in order to engender a spirit of loyalty or patriotism, but also provides support in a more immediate sense, for example, when interest-group leaders endorse legislative programs and voters select legislators at the polls. [Jewell and Patterson, 1977:10]

Solidarity of support for a political institution is usually viewed as an attribute of the constituency; however, the actions taken by an institution like a legislature can affect levels of support, while levels of public support can also affect the institution.

PROBLEM INVESTIGATION

One component in any decision-making process is an investigation of the problem and potential solutions prior to the initiation of any action. While the degree of thoroughness attending any investigation can vary considerably, some effort is usually made to examine the problem. In a legislative setting, hearings are held on most legislation and expert testimony is accepted. Further, staff and agency analysis is usually supplied. In some settings, the legislature has also developed the capacity to initiate investigation of long-range problems and future difficulties. Developing this capacity strengthens the legislature's role in making decisions.

MEASURES OF ORGANIZATIONAL PERFORMANCE

Having identified and elaborated on seven major operative goals and functions for the legislative process, we shall now outline various indicators that could be used to measure each. These indicators are intended to be *suggestive* and by no means exhaust those that could be included. By collecting data for each of these indicators, the media or interested citizens could determine in a systematic and objective manner how well a particular legislature is performing in pursuing the seven goals. Collecting and analyzing these data across time periods, for example, several sessions of a county board or a state legislature or of Congress, could also provide information regarding whether or not a legislature is performing in a better manner now as compared with two or four or six years ago. Evaluations based on such data could be utilized by a legislature to improve its performance level by adjusting the areas judged to be less than adequate—a demonstration of accountability.

POLICY RESPONSIVENESS

1. *Policy Response* considers the degree to which a policy enacted by the legislature meets and ameliorates the policy needs and demands in that policy area. Such analysis must be conducted separately for various substantive areas of policy—welfare, education, crime control and corrections, pollution control, public health, transportation, and so on.
2. *Timeliness of Policy Response* refers to the amount of time necessary for the legislature to act after a need appears to exist.
3. *Policy Congruence* is the degree to which policy decisions agree with citizen preferences as indicated by public opinion polls and other demand measures.
4. *Differential Treatment* assesses the degree to which various interest sectors and requestors of legislation obtain policy decisions favorable to their causes.

FORMAL DECISION MAKING

1. *Processing Effectiveness* is the degree to which the legislative body is able to act upon demands made of it and convert these demands into outputs. It is an evaluation of *how well* the legislature operates in a decision processing sense. The dimensions are productivity, expeditiousness and efficiency.

   (a) *Productivity* is the legislature's "conversion ratio" of output (policy decisions) to input (bills).[1]

   (b) *Expeditiousness* is the speed with which a legislature acts. For these indicators, the average number of session days required for bills to go from one decision point in the legislative process to the next might be calculated.[2]

   (c) *Efficiency* is the cost in terms of session time and dollars incurred by each chamber in taking alternative courses of action on bills. The number of days in session, the number of bills in the categories, and costs must be determined.[3]

2. *Technical Quality of Decisions Made* is the relative number of changes required in the future to correct "technical" defects in all legislation passed.

   (a) Number of technical corrections made by "statute revisor" in legislation passed during a session.

   (b) Number of clarifications of legislative intent, and so forth, made in future by legislature itself.

   (c) Number of judicial "modifications" made in future.

3. *Policy Complexity* considers the degree to which policy proposals are intricate and complicated.

   (a) Average amount of revision undertaken in statutes, and the like.

   (b) Average degree to which policy components specify action rather than leaving this for agency interpretation.

   (c) Average number of pages and subsections per bill.

OVERSIGHT OF THE ADMINISTRATION

1. *Committee Oversight* refers to the legislature's use of committees to check on administrative agencies. This includes the work of standing substantive and review committees, as well as special committees.

(a) Proportion of all committees which have taken some action on oversight (hearings, bills, resolutions, and so on.)
(b) Proportion of special committees to review work of administrative agencies.
(c) Proportion of all committee hearings devoted to agency review.
(d) Proportion of committee time devoted to agency review activities.
(e) Number of actions initiated by committees to review agency operations.
(f) Number of observed changes in agency operations as a result of legislative oversight.

2. *Legislative Staff Oversight* includes the activities of all types of legislative staff–personal, committee, and legislative agencies such as reference bureaus–in reviewing agency activities.

(a) Percentage of all bills and resolutions introduced *directing* an agency or commission to initiate some action which is contrary to current practice.
(b) Number of legislative bills and resolutions passed/Number introduced which *direct* some action by agencies.
(c) Percentage of actions *taken* by an agency in response to a legislative directive which are consequently reversed by court action or subsequent legislative directive.
(d) Number of changes made in agency operations due to legislative directives.
(e) Proportion of all members' time spent in legislative oversight.

CONSTITUENT REPRESENTATION IN POLICY FORMATION

1. *Authorization* is the degree to which legislators are given power by citizens to act in a decision-making capacity. These indicators are collected by asking constituents and legislators appropriate questions in an interview setting.

(a) Constituents' predispositions toward granting power to legislatures and legislators.
(b) Legislators' view of the adequacy of authority vested in them by constituents.
(c) Constituents' views of the adequacy of the authority vested in legislators.
(d) Constituent actions taken to reduce power given to legislators and legislatures.

2. *Responsibility* is the degree to which constituents believe and act to hold legislators liable for their actions and their policy *decisions.* Indicators of accountability are collected through questions asked of constituents and legislators or from official records.

    (a) Constituents' feelings of being able to control legislators and the legislature.
    (b) Defeat ratio and margin of electoral victory of legislators who acted counter to constituents' expressed wishes.
    (c) Reasons why incumbents left office (e.g., fear of defeat at reelection).
    (d) Reasons given for voting against incumbents.

CONSTITUENT RELATIONS, EDUCATION, AND ADVOCACY

1. *Constituent Communications* refers to the type, amount, and quality of legislators' and legislature's communications with citizens.

    (a) Proportion of responses by legislators to constituent communication (letter, telephone, and so on).
    (b) Time delay between receiving constituent communication and response.
    (c) Number of public appearances, speeches, and so forth, initiated by legislator.
    (d) Type and quality of district-wide mailings.

2. *Action on Constituent Behalf* includes the amount and type of efforts legislators take for constituents.

    (a) Percentage of legislators' time spent in case work and "errand boy" activities.
    (b) Percentage of staff time—all kinds of staff—spent in case work and "errand boy" activities.

3. *Public Outreach* refers to those legislative level activities intended to meet constituent desires for information and response.

    (a) Size and extent of legislature's public relations effort, e.g., publicized toll-free telephone numbers for securing information, explanatory booklets, public information offices.
    (b) Number and type of public notices distributed regarding legislative hearings and committee meetings.

(c) Proportion of committee hearings and investigations held away from the state capital and/or in the evening.

(d) Amount of citizen participation in committee hearings and investigations.

(e) Number of legislative reports distributed and basis used for their distribution.

(f) Average time required in filling public requests for reports and information.

SOLIDARITY BUILDING

1. *Diffuse Support* is the constituents' level of "good will" for the legislature. These indicators are measured using questions asked of constituents in standard interview settings.

   (a) Compliance with legislative actions.

   (b) Commitment to legislative process.

2. *Specific Support* is the constituents' feelings about the legislative process as a result of specific actions taken by the legislature. Information on these indicators is collected from interviews with constituents.

   (a) Constituent ratings of legislative performance.

   (b) Constituents' perceptions of benefits accrued from legislative actions.

PROBLEM INVESTIGATION

1. *Investigation by Committees* is a long-standing tradition in legislatures whereby various standing committees direct staff and hold hearings to look into problems likely to be encountered in future policy making.

   (a) Number of committee hearings devoted to long-range problems rather than specific legislation.

   (b) Amount of total staff time used to prepare position papers and so on, concerned with future problems and long-range solutions.

   (c) Proportion of all committees devoted to special-study committees having a "broad" charge to investigate long-range problems.

   (d) Proportion of legislative council time spent in investigating long-range problems.

   (e) Number of policy proposals generated and passed from various long-range investigations.

In assessing a legislature's performance for these seven functions, one must recognize that individuals probably place different values on each function and may actually place little, if any, value on one or more of them. Further, some functions may actually conflict with one another, so that maximizing one may minimize another. At the extreme, some functions may be incompatible so that productivity may conflict with solidarity building or with representation, and so on.

STRATEGIES FOR INCREASING ACCOUNTABILITY IN LEGISLATIVE BODIES

We have suggested an approach for measuring levels of legislative performance, and now turn our attention to strategies which may be utilized by the legislature, citizens, or other groups for improving accountability. The strategies to be discussed here are suggestions collected from various sources and are presented to stimulate discussion.

CONTINUED STRUCTURAL/PROCESS/PERSONNEL CHANGES

One obvious strategy is to continue the open process/structural change approach which emphasizes improving the visibility of and understanding of already-in-place structures and processes—a la Common Cause or Legis 50. Some examples of this approach, which have been implemented, are opening up the process by which legislative decisions are made, fostering understanding of the legislative process, replacing nonresponsive public officials through the election process, and increasing political party responsibility. Strides have been made through individual and group efforts in these directions; however, the question remains: "Will these efforts produce greater accountability?"

At best this remains an open question, at least as far as institutional accountability is concerned. Certainly the fragmentary evidence presented above suggests that state legislatures ranking relatively higher on accountability actually rank lower on several measures of performance, trust, and confidence. This certainly raises questions regarding a continuation of this effort as the *sole* and perhaps even as the *major device* for increasing accountability.

INITIATIVE/REFERENDUM/RECALL

One legacy in our political system from the Progressive Reform Movement is the use of initiative, referendum, and recall. These devices were implemented to provide citizens with opportunities for *direct involvement* in policy decision making thus supplanting the need for other accountability devices. Robert Weissberg (1976:67-68) has argued that five conditions should exist in order for initiative and referendum to be effective mechanisms for assuring direct democracy and accountability.

> First, voters would have at least some opportunity to decide on substantial issues framed in intelligible terms. Second, voting would be extensive and representative of all citizens, not just a small minority. Third, a fair degree of competence would be displayed by voters on each issue and all the decisions taken in the aggregate. Fourth, the availability of direct legislation would, in some sense, have a political policy impact apart from merely adding more issues to the electoral campaigns. Finally, the opportunity to participate successfully in direct legislation would be reasonably available to a variety of groups and interests.

In reviewing evidence for each of these requirements from those states having initiative and referendum provisions, Weissberg concludes that these conditions have been inconsistently satisfied. While many substantial issues have been placed on the ballot for direct citizen vote, many more have not. Further, a large number of relatively minor issues have also appeared on the ballot. Second, voter turnout for initiative and referendum issues has generally been lower than for general elections. Further, Charles M. Price (1975:257) analyzed differences in the levels of accountability and overall performance levels using the Legis 50 rankings, for high initiative, low initiative, and noninitiative states. Based upon measures of statistical association, he concluded that

> there is little difference in the quality of the legislature (overall, or accountable) between high-use and low-use initiative states or between initiative and non-initiative states. Hence, the conventional wisdom view that frequent initiative use denotes legislative ineptness or failure can, at least, be questioned.

In spite of the availability for recalling state-wide officials in 12 states and local officials in 28, it is used relatively infrequently. Only two occasions exist where recall was successful on a state-wide basis—North Dakota and Oregon—and in one of these the governor, after being recalled, was elected to the United States Senate (Levitt and Feldbaum, 1973:252; Maddox and Fuquay, 1966:333). On the local level, no precise information is available regarding the frequency of use or success rate. One study determined that during the first 25 years, only 72 recall elections were held in 952 cities (Maddox and Fuquay, 1966:333). While the ever-present threat of recall may lead a public official to be more sensitive to citizen demands and more cautious in his official actions, recall seems to be more of a potential source for achieving accountability than a real source.

> There is a trend away from the recall, and it has never been widely used. Few new adoptions of it have taken place since 1920 . . . But the question of the use of the recall is no longer very important. [Adrian, 1967:172]

Although initiative, referendum, and recall are all currently available to varying degrees in our political system and would appear to be devices that could be used to achieve greater accountability, the future prospects seem dim as use is not very widespread. Public interest in these devices seems to be waning from a level which appears never to have been very high. A major campaign to revive and expand use of these devices would seem to be a prerequisite to any serious treatment of them as devices for improving accountability; and even then, it remains problematic, according to the data cited above, whether initiative, referendum, and recall could ever be effective in securing greater accountability.

INDEPENDENT OVERVIEW

A third strategy that might be followed emphasizes the collection and dissemination of information on performance levels of legislative bodies by independent agencies. The goal is to have more electoral decisions and public pressure based on overall performance data. One format, suggested by E. Terrance Jones (1977) involves the media periodically distributing legislative performance data. As Jones points out, the newspapers have developed a system whereby their readers can be informed through rigorous, systematic, and quantitative data about the ongoing as well as seasonal performance of virtually any kind of sports team—won-lost record, winning percentage, completed passes, and so on. If this can be done for sports, it can also be done for other spheres of activity like legislatures and government in general. Certainly, a great amount of work would have to be done to develop and refine measures and to present them in clear and understandable form to the public, but from one perspective, this approach is an extension and modification of what many newspapers already do. Numerous articles are printed summarizing the legislature's actions on one or a few bills in most sessions of Congress or a state legislature. "Box-scores" of voting records, absences, and group support scores are commonly reprinted in newspapers. "Insider" reports of legislator behavior both inside and outside the legislature are widespread. What is unique about the Jones-based proposal is that a much wider range of legislative activity would be observed, greater attention would be paid to defining what is to be used, and a greater amount of information would be used.

Several drawbacks characterize this proposal: it could be very costly to collect and analyze these data for the many legislative bodies in our political system; great harm could be done if objective, accurate, and valid data were not collected; and no assurances exist that citizens would make use of such evaluations to pursue greater institutional accountability. Yet, this proposal would seem to offer some promising developments for citizens and groups concerned about increasing accountability.

LEGISLATIVE INSPECTOR GENERAL

Probably the most extreme step which could be taken in order to achieve greater accountability would be to create an independent agency to secure it. Several political systems, including our own, have in varying degrees experimented with this proposal. Certainly various audit and program review agencies currently perform in this manner. An independent agency, called perhaps the "Legislative Inspector General," could be created and charged with the responsibility of monitoring the operations of the legislative branch of government. This agency might collect performance data such as those suggested herein, and monitor changing levels in these measures across time. Whenever certain measures indicate a significant drop in performance, the Legislative Inspector General could be authorized to seek remedies including a recall vote or a referendum on the actions, a plebiscite on various issues, or a legal action to seek a judicial remedy. This strategy would institutionalize accountability efforts in the hands of an independent agency.

Probably the greatest drawback for this alternative is the possibility of creating yet another bureaucracy. Taken to its logical extreme, one might conclude that it is necessary to have an "Inspector General" review the performance of the "Legislative Inspector General," and so on. Further, is the notion of creating institutionalized review agencies with the power to initiate actions to secure accountability compatible with our political system? Certainly, this proposal raises philosophical as well as practical questions.

While each of these four strategies for increasing levels of legislative accountability poses some promise to the citizen concerned with maintaining responsible public officials, each also poses potential problems, some of which may be greater than those inherent in our current system of accountability. Nevertheless, the time appears to have come when serious attention on the part of public officials and citizens must be directed toward resolving the issue of accountability.

SUMMARY

This chapter has addressed the issue of accountability in legislative bodies. While many questions remain to be answered, we directed attention to some critical issues inherent in studying accountability and suggested some answers. For example, we argued that

(1) Institutional accountability for the entire legislative body is important and should be pursued with the same vigor and resolution as is individual level accountability.

(2) In order to pursue accountability, individual citizens and groups should focus on all of the aspects of institutional and personal performance which are germane for assessing the policy-making role of the legislature. To evaluate accountability *only* in terms of the policy decisions overlooks many vital aspects of policy making and institutional responsibility.

(3) Objective, valid, and useful performance measures for a legislative body can be developed if one uses the approach taken in social psychology, organizational theory, and management science to specify organizational goals and their indicators. Collecting information on these goals and charting measures of performance across time would give to public officials, citizens, and public-minded organizations a means for assessing and reviewing the legislature's operations and would provide a firm evidential base for establishing institutional and individual accountability.

(4) There are four differing strategies for increasing accountability—continued structural/process/personnel changes, initiative/referendum/recall, independent overview, and a legislative inspector general. Each of these offers some promise for increasing levels of accountability but each also has inherent problems. Different legislatures—and their political systems, local, state, and national—are likely to experiment with each of these strategies as well as with new ones that may emerge. Regardless of the choice of strategies, the essential element is for some action to be initiated by academics, public officials, citizens, and the media. Much work needs to be accomplished in studying accountability—especially regarding the five questions posed at the outset of this chapter—and its implementation, in setting into place official policies and strategies for assuring increased accountability to the public, in educating citizens regarding their role in achieving accountability, and in publicizing information prerequisite for achieving accountability.

NOTES

1. Productivity measures suggested elsewhere could be: Number of bills debated and amended/Number of bills introduced; Number of bills passed/Number of bills debated and amended; Number of bills passed/Number of bills introduced (Hedlund and Hamm, 1977).

2. Expeditiousness measures could be: Days from bill introduction to final disposition; Number of days behind in schedule; Days from committee report to scheduling the bill for debate and amendment; Days from scheduling bills for debate and amendment to actual debate and amendment; Days from final review (engrossment) to final disposition (final vote, tabling, and so on; Hedlund and Hamm, 1977).

3. Efficiency measures could be: Session days/Bills introduced; Session days/Bills passed; Legislative costs/Bills introduced; Legislative costs/Bills passed (Hedlund and Hamm, 1977).

REFERENCES

ADRIAN, C. R. (1967). State and local governments (2nd ed.). New York: McGraw-Hill.

BLACK, M., KOVENOCK, D. M., and REYNOLDS, W. C. (1974). Political attitudes in the nation and the states. Chapel Hill, N.C.: Institute for Research in the Social Sciences.

BLAIR, G. S. (1967). American legislatures: Structure and process. New York: Harper & Row.

BURNS, J. (1971). The sometimes governments. New York: Bantam Books.

Citizens Conference on State Legislatures (1971). State legislatures: An evaluation of their effectiveness. New York: Praeger.

EULAU, H. and PREWITT, K. (1973). Labyrinths of democracy: Adaptations, linkages, representation, and policies in urban politics. Indianapolis: Bobbs-Merrill.

FENNO, R. F., JR. (1975). "If, as Ralph Nader says, Congress is 'the broken branch,' how come we love our congressmen so much?" Pp. 277-287 in N. J. Ornstein (ed.), Congress in change: Evolution and reform. New York: Praeger.

FRANCIS, W. L. (1972). Formal models of American politics: An introduction. New York: Harper & Row.

HEDLUND, R. D., and HAMM, K. E. (1977). "Institutional development and legislative effectiveness: Rules changes in the Wisconsin Assembly." Pp. 173-213 in A. I. Baaklini and J. J. Heaphey (eds.), Comparative legislative reforms and innovations. Albany, N.Y.: State University of New York Press.

--- (forthcoming). "Institutional innovation and performance effectiveness in public policy making." In Leroy N. Rieselbach (ed.), Legislative reform: The policy impact. Lexington, Mass.: Lexington-Heath.

IPPOLITO, D. S., WALKER, T. G., and KOLSON, K. L. (1976). Public opinion and responsible democracy. Englewood Cliffs, N.J.: Prentice-Hall.

JEWELL, M. E., and PATTERSON, S. C. (1977). The legislative process in the United States (3rd ed.). New York: Random House.

JONES, E. T. (1977). Presentation at Urban Research Center colloquium, University of Wisconsin-Milwaukee, February 14.

KATZ, O. and KAHN, R. L. (1966). The social psychology of organizations. New York: Wiley.

KEEFE, W. J., and OGUL, M. S. (1977). The American legislative process: Congress and the states (4th ed.). Englewood Cliffs, N.J.: Prentice-Hall.

LEVITT, M. J., and FELDBAUM, E. G. (1973). State and local governments and politics. Hinsdale, Ill.: Dryden.

LOEWENSTEIN, K. (1957). Political power and the governmental process. Chicago: University of Chicago Press.

MADDOX, R. W., and FUQUAY, R. F. (1966). State and local government. New York: Van Nostrand.

PATTERSON, S. C. (1976). "American state legislatures and public policy." Pp. 139-195 in Herbert Jacob and Kenneth L. Vines (eds.), Politics in the American states. Boston: Little, Brown.

PITKIN, H. F. (1967). The concept of representation. Berkeley: University of California Press.

PREWITT, K. (1970). The recruitment of political leaders. Indianapolis: Bobbs-Merrill.

PRICE, C. M. (1975). "The initiative: A comparative state analysis and reassessment of a western phenomenon." Western Political Quarterly, 28 (June):243-262.

PRICE, J. L. (1968). Organizational effectiveness: An inventory of propositions. Homewood, Ill.: Irwin Press.

RIESELBACH, L. N. (1975). "Congress: After Watergate, what?" Pp. 66-118 in Leroy N. Rieselbach (ed.), People vs. government: The responsiveness of American institutions. Bloomington: University of Indiana Press.

SALOMA, J. S., III (1969). Congress and the new politics. Boston: Little, Brown.

VAN DER SLIK, J. R. (1977). American legislative process. New York: Crowell.

WEISSBERG, R. (1976). Public opinion and popular government. Englewood Cliffs, N.J.: Prentice-Hall.

4

Accountability in the Administrative State

DAVID H. ROSENBLOOM

☐ THE EMERGENCE OF THE FULL-FLEDGED ADMINISTRATIVE STATE during the twentieth century has created many barriers to governmental accountability. Indeed, some of these appear to be so insuperable that it could reasonably be held that the development of the administrative state as we know it actually precludes the possibility of maintaining mechanisms for accountability that adequately meet the needs of a democratic polity. This chapter will address the character of the administrative state, the barriers it presents to accountability, and some alternative ways of dealing with the overall political problem of assuring responsive government despite the agglomeration of power in the hands of public bureaucrats.

THE ADMINISTRATIVE STATE

Although one can trace the development of the modern administrative state from the 1660s (Barker, 1966), it has only recently reached fruition. The administrative state is characterized by a large and increasing governmental role in the social and economic life of the political community. As the government takes on additional and more complex functions, these are implemented by administrative organizations. In the early days of the administrative state, the relationship between political power and administration seemed clear. Political authorities established the policy goals of the state, whereas administrators chose the means of obtaining these objectives. However, as the size and complexity of the administrative component of modern government grew, administrators began to emerge as a major power center in their own right.

Today, it is not excessive to follow Wallace Sayre (1965:2) in describing their power as follows:

> The staffs of the executive branch agencies have come to exercise an important share of the initiative, formulation, the bargaining and the deciding in the process by which governmental decisions are taken. They are widely acknowledged to be the leading "experts" as to the facts upon which issues are to be settled; they are often permitted to identify authoritatively the broad alternatives available as solutions; and they frequently are allowed to fix the vocabulary of the formal decision. These powers are shared and used by the career staffs in an environment of struggle and competition for influence. . . .
>
> Great power also belongs naturally to those who carry out decisions of public policy. In this stage the career staffs have had a paramount role. The choice of means, the pace and tone of governmental performance, reside largely in the hands of the [civil] service. Constraints are present, and most of these uses of discretion by the career staffs are subject to bargaining with other participants, but the civil servants have a position of distinct advantage in determining how public policies are executed.

It is important to recognize in this connection that the "choices of means" hardly confines administrative power to matters of detail or minor importance. A great deal of legislation and executive direction delegates authority to administrators subject only to very broad standards, such as that the powers granted be exercised in the public interest. This enables administrators not only to choose from among a broad variety of means, but also to develop their own particular view of the public interest or the other overall objectives being sought. But often the choice of means under these conditions is as much a political decision as selection of the ends themselves. To take but one contemporary example, the use of the term "affirmative action" to require the utilization of goals or quotas for the employment of members of minority groups and women, as covered by the Civil Rights Act of 1964 and the Equal Employment Opportunity Act of 1972, is purely an administrative creation. Yet it is the goals or quotas rather than the end of equal opportunity which is now most politically controversial. It is for these reasons that some have contended that the administrative state would come to a "standstill" if public administrators did only what they were told (Storing, 1964:152).

The rise of administrative power does not bode well for the maintenance of accountability along democratic lines. Traditional democratic theory approaches the problem as follows:

> power emanates from the people and is to be exercised in trust for the people. Within the government each level of executive authority is accountable to the next, running on up to the President or the Cabinet. The executive authority as a whole is accountable to the Congress or Parliament, which is assisted in its surveillance of expenditures by an independent audit agency. Officials are required to submit themselves to periodic elections as a retrospective evaluation of their performances and to receive a new mandate from the people. [Smith and Hague, 1971: 26-27]

But as the administrative state grew in size and complexity, this approach became highly dubious. Today, "accountability gets lost in the shuffle somewhere in the middle ranges of the bureaucracy" (Smith and Hague, 1971:27), because the elected and politically appointed officials are not firmly in charge. Hence, the rise of the administrative state leaves us with the question of "How does one square a permanent civil service–which neither the people by their vote nor their representatives by their appointments can replace–with the principle of government 'by the people'?" (Mosher, 1968:5). From the perspectives not only of accountability, but also of democracy, this, then, is the central issue posed by the emergence of the administrative state. In order to understand it better, we must first explore the nature of administrative power.

BUREAUCRACY

The organization of administrative power in the modern state is overwhelmingly bureaucratic. Although the term "bureaucracy" is often used as a synonym for inefficiency, red tape, and government agencies, it has a more technical and useful meaning as well. "Bureaucracy" as commonly conceptualized in the social sciences today consists of at least the following central characteristics: hierarchy, specialization, formalization, a personnel system ostensibly based on merit and seniority, large size, and the production of "output" that cannot be evaluated in free market quid pro quo transactions. These characteristics afford both individual bureaucrats and bureaucratic organizations as a whole with a strong shield against meaningful direction by outsiders, including political executives and legislatures. Hence, the very nature of bureaucracy limits its accountability in several respects.

The tap root of bureaucratic power, perhaps, is specialization. Specialization enables individuals and organizational units to acquire great expertise in relatively narrow areas. This knowledge may be exceedingly technical, as is often the case in military procurement or health programs, or it may be more general as is often true in such areas as equal opportunity and social welfare. In either case, however, as a general rule it can only be disregarded with peril. As Max Weber (1958:229) expressed it in his classic statement on bureaucracy,

> The ruled, for their part, cannot dispense with or replace the bureaucratic apparatus of authority once it exists. For this bureaucracy rests upon expert training, a functional specialization of work, and an attitude set for habitual and virtuoso-like mastery of single yet methodically integrated functions. If the official stops working, or if his work is forcefully interrupted, chaos results, and it is difficult to improvise replacements from among the governed who are fit to master such chaos. This holds for public administration as well as for private economic management. More and more the material fate of the masses depends upon the steady and correct functioning of the increasingly bureaucratic organizations of private capitalism. The idea of eliminating these organizations becomes more and more utopian.

Moreover, "Under normal conditions, the power position of a fully developed bureaucracy is always overtowering. The 'political master' finds himself in the position of the 'dilettante' who stands opposite the 'expert' facing the trained official who stands within the management of administration" (Weber, 1958:232). The modern, complex polity relies upon expertise, and public bureaucracy is among the state's chief suppliers of this valued commodity.

But what is true for the "political master," is also true to a lesser extent for those who occupy the top ranks of the bureaucratic organization's hierarchy. The purpose of hierarchy is to coordinate the specialists for "what has been taken apart must be put together again. A high degree of specialization creates a need for a complex system of coordination" (Blau and Meyer, 1971:8). Those in the top ranks, however, are not supposed to *duplicate* the specializations of their subordinates, even if it were possible to do so. Consequently, they, too, must rely—sometimes almost blindly—upon their subordinates for information concerning what can and cannot be done, how, and why. This means that while hierarchy is formally a method of controlling subordinates, in reality bureaucratic subordinates are often assured a good deal of independence by their specialized expertise. Indeed, Victor Thompson (1961:6) has argued that "the most symptomatic characteristic of modern bureaucracy is the growing imbalance between ability and [hierarchical] authority." Cabinet members in the United States present a good example of this. Some, such as the head of the Department of Health, Education and Welfare, are given a virtually impossible task. While theoretically responsible for what occurs within their departments, it is clear that they must rely very heavily upon subordinates because no single individual could even hope to master all the information and knowledge necessary to fulfill the organization's missions. A former Undersecretary of the Department of Commerce summed up the situation in the following terms:

> In a real sense, delegation of authority to the operating manager of an entirely unfamiliar field means that the Secretary serves the bureau chiefs rather than vice versa.
>
> [The Secretary's] judgment on budget items is, of course, the most important decision he will make in his term in office and is the decision he is usually least well-equipped to make intelligently.
>
> The traditional answer of the busy executive to excessive workload—delegation of authority—is often a high-risk business in a political organization. [Bartlett and Jones, 1974:63, 64, 63]

Despite the inherent limitations upon the real authority of those at the top of hierarchical organizations, they might nevertheless exercise power to hold subordinates accountable for their actions, were it not for the nature of bureaucratic personnel systems.

Bureaucratic personnel systems are based on "achievement oriented" rather than "ascriptive" criteria. Thus, individuals are appointed on the basis of their ability to perform the work involved, rather than with reference to their political or social characteristics. Promotions are generally conditioned by seniority as well. In the public sector in the United States the standard mechanism for ascertaining ability is merit selection, which often, though not always involves some sort of written or performance examination. Typically, the hiring official is allowed to offer an appointment to an individual who is among the top three candidates in terms of examination score. This in itself has substantial consequences for accountability. It means that since the hiring official is not free to make selections according to his or her wishes, the dismissal of an employee will not in itself guarantee better performance from the successor. The situation is complicated immensely, however, by restrictions on the right to dismiss employees.

When the merit system was originally introduced into the United States in the 1880s, many of the civil service reformers wanted to leave dismissals unregulated. They feared that "If it were necessary to establish unfitness or indolence, . . . by such proof as would be accepted in a court of law, sentence would seldom be pronounced, even against notorious delinquents" (Rosenbloom, 1971:87-88). However, the prevailing tendency in public bureaucracies has been, as Weber (1958:202) noted, that "normally, the position of the official is held for life" and by the early 1900s civil service protectionism was on the rise. On the federal level, by 1912 the Lloyd-LaFollette Act provided civil servants with some protection from dismissal "except for such cause as will promote the efficiency of [the] service." Subsequently, first veteran preference eligibles and then all other nonprobationary employees were given a right to a hearing in dismissal and some other adverse actions. Although the Supreme Court's decision in *Arnett* v. *Kennedy* (1974) created the possibility that these hearings might be after the fact, the general tenor of the personnel system remains highly protective of the employee. As Louis Gawthrop (1969:147) points out, the supporters of the current personnel approach

> seek to maximize individual security and protection against arbitrary personnel decisions. They agree that effective and efficient administrative actions can only be realized within a relatively stable organizational environment in which individual anxieties and tensions generated by feelings of occupational insecurity have been significantly eliminated. If one can remove the causes of such anxiety–threatened dismissals, reductions in rank, arbitrary reward allocations–then personnel security can be realized, administrative continuity develops, and operating efficiency increases.

As a result of this general personnel philosophy, and other factors including judicial decisions and the public sector labor movement which we shall discuss below, public employees at most levels enjoy a great deal of job security. Adverse actions are possible, but they often resemble lawsuits and managers are generally reluctant to engage in them. Indeed, an agency may prefer to go through a reorganization or a reduction-in-force rather than an outright dismissal in order to rid itself of unwanted employees.

The job tenure afforded by public personnel systems provides administrators with a great deal of independence. This in turn limits their accountability to political superiors. Public employees sometimes resist direction from politically appointed or elected officials, often viewing them as little more than transitory figureheads. When this occurs there is relatively little that these political officials can do about the situation except to offer the bureaucrats something in return for their cooperation or otherwise try to win their support. Such a course, however, has its costs in terms of accountability. As one member of the Executive Office of the President put it, "after six to twelve months even the political appointees in the departments get captured and taken in by the agencies" (Cronin, 1975: 168). Perhaps none has improved on the lament of James J. Davis, Secretary of Labor under President Harding, that "the simple fact is that I am powerless to enforce changes which I desire because I am powerless to put in charge of these places individuals in sympathy with such changed policies" (Rosenbloom, 1971:240). In short, the bureaucrats wielding the power of the administrative state, are relatively invulnerable to personal sanctions applied by the political authorities who seek to control the government's administrative component.

Formalization and the related bureaucratic impersonality also make it difficult to obtain accountability in the administrative state. Formalization refers to the bureaucratic penchant for communication in writing. Because the structure is intended to continue even as individual personnel move in and out of a bureaucracy, communication takes the form of memos and directives written from one position or unit to another. In addition, many aspects of organizational life are set forth in a formal written fashion. In the federal bureaucracy, for instance, there is a formal description of the responsibilities of every position. While the principal object of formalization is clarity, the desire to be precise tends to create a special variant of the language. This tendency is reinforced by the nature of bureaucratic specialization. The net result is very often a language that protects bureaucratic power and militates against accountability.

Ralph Hummel (1977) has addressed this feature of bureaucracy in some detail. Starting from the premise that bureaucratic language is one-way communication, that is the bureaucrats or superordinates talk and others listen, he argues that administrators "are not free to redefine their vocabularies in a two-way exchange" (1977:145), which is the normal course of communication in society at large. Consequently,

> the function of bureaucratese is fundamentally to make outsiders powerless. . . . Bureaucratic specialized language is specifically designed to insulate functionaries from clients, to empower them not to have to listen, unless the client first learns the language. For a client who has learned the language is a client who has accepted the bureaucrat's values. Language defines both what problems we can conceive of and what solutions we can think of. Once a client uses the bureaucracy's language, the bureaucrat may be assured that no solutions contrary to his interests and power will emerge from the dialogue. [Hummel, 1977:147]

The same, of course, is true with respect to political authorities trying to interact effectively with a bureaucracy.

Bureaucratic impersonality is closely aligned with formalization. Impersonality refers to the separation of the individual as a social being from his or her existence in the organization. Sometimes referred to as "dehumanization," it has been called bureaucracy's "special virtue" because it eliminates emotions, personal biases, and idiosyncracies from the performance of individual bureaucrats (Weber, 1958:216). As a result, the organization takes on a life of its own, independent of that of the individuals who staff it. Thus, bureaucratic agencies may develop distinctive "cultures" (Seldman, 1970) and policy outlooks. This, in turn, poses a problem for accountability because it becomes necessary to hold an organization, rather than one or a group of individuals responsible for the agency's actions. But how does one hold a *structure* or other intangible accountable? As Hummel (1977:147) writes, "if there is anything to talk back to at all, it is a *structure,* not a person." Reorganization is the preferred approach for dealing with this problem, but it, too, is fraught with difficulties which will be discussed below.

The large size of many governmental bureaucracies further militates against accountability. Although many questions pertaining to the impact of organizational size upon organizational behavior have yet to be fully answered (Hall, 1972), there is little doubt that large size reinforces the need for hierarchy, formalization, and impersonality. Furthermore, it may reduce accountability by simply making an organization too big to manage effectively. For example, on the face of it, this appears to be the case with both the United States Department of Defense and the Post Office. In addition, large size can enable employees to retain a relatively high degree of anonymity, thereby making accountability more difficult.

The absence of free external markets for the evaluation of the output of governmental bureaucracies presents a very substantial barrier to accountability. Put simply, when all the political rhetoric about bureaucratic inefficiency is put aside, it is very difficult to know if a bureaucracy is actually doing its job well. This is true even when program objectives are relatively clear, which is decidedly not always the case. For instance, how do we know if the Department of Defense is performing efficiently? Perhaps we could have the same defense and deterrent capacity despite a reduction in defense spending and employment, but short of incurring a catastrophe, there is no way of knowing when the critical point will be reached. When we turn to bureaucratic programs designed to provide equitable regulation and justice, the issue is even further clouded because here it would be difficult even to develop agreement on the nature of meaningful indicators of bureaucratic performance.

But when the issue is posed in these terms it is at least relatively straightforward. However, in the real world of bureaucratic politics it is not always possible to specify program objectives, much less measure progress toward them. American bureaucracies are characterized by pluralistic organization and politics. Overlaps and redundancies abound. For example, in 1971 President Nixon claimed that there were nine departments and 20 agencies involved in health; three departments involved in water resource management; two departments and four agencies involved in the management of public lands; three departments and six agencies administered various federal recreational programs; seven agencies provided assistance for water and sewer systems; six departments collected similar economic information; and seven departments were concerned with international trade (Nathan, 1975:137). This situation has several important consequences.

First, by definition, pluralistic organization fragments political power. As a result, administrators must develop and maintain support for their agencies and programs. As Norton Long (1965:16) wrote:

> It is clear that the American system of politics does not generate enough power at any focal point of leadership to provide the conditions for an even partially successful divorce of politics from administration. Subordinates cannot depend on the formal chain of command to deliver enough political power to permit them to do their jobs. Accordingly they must supplement the resources available through the hierarchy with those they can muster on their own.

Creating support, however, places a heavy premium on building a consensus favoring various bureaucratic programs. Yet, in many cases, nothing is more likely to obscure the establishment or program goals and priorities more than the attempt to develop a broad base of support. This should be abundantly clear from the nature of political platforms. The only difference is that while we tend to accept unclarity and irrationality in politics, many are wont to reject it in public administration.

Be that as it may, there are several existing bureaucratic programs and agencies whose goals are self-contradictory and/or obscure. Let the doubtful look at the Government Organization Manual's (1972/1973:251) description of the mission of the Department of the Interior:

> In formulating and administering programs for the management, conservation, and development of natural resources, the Department pursues the following objectives: the encouragement of efficient use; the improvement of the quality of the environment; the assurance of adequate resource development in order to meet the requirements of national security and an expanding national economy; the maintenance of productive capacity for future generations; the promotion of an equitable distribution of benefits from nationally-owned resources; the discouragement of wasteful exploitation; the maximum use of recreational areas; and the orderly incorporation of Indian and Alaska Native people into our national life by creating conditions which will advance their social and economic adjustment.

It is this political milieu that brings farmers, supermarkets, food processors, and advocates of the poor together in favor of a Food Stamp program within the Department of Agriculture and both advocates and enemies of equal opportunity together in support of more program responsibility for the already overburdened Equal Employment Opportunity Commission (Rosenbloom, 1977a).

In short, the major objective of many agencies and programs is to develop and maintain a political consensus for their continued existence. In order to do so, program goals and priorities may be obscured. Likewise, they may also be purely symbolic. How then, can the performance of bureaucratic agencies be evaluated by outsiders in an effort to establish accountability? This question is hardly inconsequential, for despite the current fetish with policy evaluation (Kramer, 1975), it is clear that so-called "rational" budgeting techniques such as PPBS and Zero Base Budgeting may meet their demise precisely over the inability of anyone to define bureaucratic objectives with sufficient clarity. Without knowing what an agency is supposed to do, much less how well it is doing it, it becomes impossible to hold it accountable for its program activities in any meaningful fashion. Indeed, even where goals are relatively specific, nonsymbolic, and concerned with more than the maintenance of the status quo, accountability is difficult to develop because our evaluation techniques remain relatively primitive. As Thomas Morehouse (1972:870) writes:

> Current indications are that evaluation research has so far failed to deliver. Assessments of program evaluation research work at the federal level are in general agreement not only that most of the work has been unsuccessful, but also that even the uncommon instances of technically "successful" evaluation research have failed to affect program policy making or administration in any significant way.

Perhaps accountability will be better served if these techniques are improved in the future (Nachmias, 1978).

Our discussion of the bureaucratic organization of administrative power indicates that the central features of bureaucracy militate very strongly against maintaining accountability in the administrative state. Administrators, agencies, and programs are heavily insulated from control by elected and politically appointed officials. In the United States this normal condition of the administrative state is further compounded by the constitutional status of public employees and the current directions of the public sector labor movement.

PUBLIC EMPLOYEES' CONSTITUTIONAL RIGHTS AND ACCOUNTABILITY[1]

The protective nature of public personnel systems in the United States has been augmented in recent years by the development of judicial doctrines limiting infringements by governments, in their roles as employers, upon the constitutional rights of their employees. These doctrines have almost completely reversed the approach which dominated this constitutional area from the founding period until the 1950s. As a result, it has become constitutionally difficult and often impossible to take adverse actions against public employees as a result of their speech, beliefs, or off-the-job behavior. Moreover, due process of law must now be accorded in several types of dismissals. These developments have enabled public employees to engage in behavior in opposition to elected and politically appointed officials with a reduced fear of reprisal. Consequently, as freedom has been gained an additional measure of accountability to these officials has been lost.

Historically, there have been four main approaches to the constitutional status of public employees (Rosenbloom 1971, Rosenbloom and Gille, 1975). The first, generally called the "doctrine of privilege," rested on the presumption that because holding a government job is a voluntarily accepted privilege rather than a right or coerced obligation, governments could place any restrictions they saw fit upon public employees, including those infringing upon their ordinarily held constitutional rights as citizens. For example, in *Bailey* v. *Richardson* (1950:58-59), a case arising out of the federal loyalty-security program, it was held, and subsequently affirmed by an equally divided Supreme Court (1951), that "due process of law is not applicable unless one is being deprived of something to which he has a right." Consequently, "the plain hard fact is that so far as the Constitution is concerned there is no prohibition against the dismissal of Government employees because of their political beliefs, activities or affiliations." Thus, the government had great leeway in dealing with its employees who could be (and were) dismissed for such things as favoring racial integration, reading Tom Paine or even the *New York Times,* and not attending church services, as well as for engaging in a host of unconventional or nonconformist activities (Rosenbloom, 1971).

Whatever the inherent constitutional logic of this approach, it could not withstand the vast increases in public employment of the 1950s and 1960s, the emergence of new concepts of civil rights and civil liberties under the Warren Court, and the growing dependence of citizens outside public employment upon such governmental "privileges" as welfare payments. During the 1950s and 1960s, the doctrine of privilege was replaced by what has been called the "doctrine of substantial interest" (Rosenbloom, 1971). This approach starts "with the premise that a state [or the federal government] cannot condition an individual's privilege of public employment on his non-participation in conduct which, under the Constitution, is protected from direct interference by the state" (*Gilmore* v. *James*, 1967:91). Moreover, "when there is a substantial interest, other than employment by the state, involved in the discharge of a public employee, he can be removed neither on arbitrary grounds nor without a procedure calculated to determine whether legitimate grounds do exist" (*Birnbaum* v. *Trussell,* 1966:678). Although largely developed in the lower courts, the doctrine of substantial interest was embraced by the Supreme Court in cases such as *Sugarman* v. *Dougall* (1973:644) in which it stated that "this court now has rejected the concept that constitutional rights turn upon whether a governmental benefit is characterized as a 'right' or as a 'privilege.' " Elsewhere, it called such a distinction "wooden" and rejected it "fully and finally" (*Board of Regents* v. *Roth*, 1972:571). As the constitutional rights of public employees began to be equivalent to those of ordinary citizens, many civil servants began to engage in greater political activity.

Once largely free of the fear of reprisal, public employees began to speak out on the major issues of the day, including those with which they had special familiarity as a result of their jobs. Thus, in the federal service there were several antiwar petitions and statements issued by employees of the various departments, including those of State and Defense. Civil servants in the Department of Justice and others spoke out in the area of civil rights policy. Sometimes these activities took the form of protests, and on at least one occasion a department head was literally chased by some 300 employees seeking to present him with a petition concerning their grievances and demands for the furtherance of racial equality (*Washington Post,* 1970:B1). Such activities gave rise to a variety of more or less formal employee organizations dedicated to changing the nature of both public employment and the society at large.

One such group, Federal Employees for a Democratic Society (FEDS), stressed the desirability of participatory bureaucracy in which hierarchy would be reduced to a minimum, rank-and-file employees would participate in the making of public policy to a greater extent, and employees would be free to refuse to perform work that violated their consciences (Hershey, 1973:51-63). Although less well documented, similar activities took place at the state and local levels as well, and if anything, the "new militancy" (Posey, 1968) of public employees in the 1960s was most important in urban settings. The New York City teachers' strike of 1968 serves as an excellent example of a struggle for control over public policy. There, the central question was who should control the schools; elsewhere, similar matters arose with reference to almost the whole gamut of state and local functions. As protest, disruption, and dissent in the public service became more pronounced, some efforts were made by political authorities and public managers to stem its tide. However, under the doctrine of substantial interest, most of these activities were protected, and dismissal became highly impracticable when an employee was willing to take his or her case to court.

Understandably, elected and politically appointed public officials were often more than a little disenchanted with the effects of the doctrine of substantial interest on accountability. One of President Nixon's taped conversations with George Schultz, Director of the Office of Management and Budget in 1971, is revealing in this regard:

> You've got to get us some discipline, George. You've got to get it, and the only way you get it, is when a bureaucrat thumbs his nose, we're going to get him.... They've got to know, that if they do it, something's going to happen to them. Where anything can happen. I know the Civil Service pressure. But, you can do a lot there, too. There are many unpleasant places where Civil Service people can be sent. We just don't have any discipline in government. That's our trouble. [*New York Times,* 1974:14]

Nixon sought to rectify the situation by creating the "administrative presidency" (Nathan, 1975) and attempting to circumvent the merit system through various abuses (U.S. Civil Service Commission, 1976). His appointments to the Supreme Court, however, took a different tack.

The next approach to the constitutional status of public employees was largely an outgrowth of the split between the liberal and Nixon appointees on the Court. This division prevented achieving a high degree of agreement on approaches which would determine the direction of the law for years to come. Majorities could generally be formed only with reference to a specific set of facts and circumstances, and broad, sweeping generalizations were consequently precluded. Hence the emergence of what can be called the "idiographic" approach, which was devoid of doctrinal, across-the-board judicial pronouncements, but rather treated each case individually on its own merits.

The tendency toward a case-by-case nondoctrinal resolution of issues was further enhanced by the requirement of the doctrine of substantial interest that the courts weigh the nature of the injury to the individual against the claimed benefit to the state in an effort to determine whether substantive constitutional rights had been violated and in order to decide what procedural safeguards, if any, had to be made available when adverse actions were taken against public employees. Thus, in several cases the Supreme Court and the rest of the federal judiciary required that, in applying general personnel regulations and principles to individual employees, public employers address the specific sets of facts and circumstances involved. For example, under this approach, it was possible to ban aliens from some, but not all public service positions; race might be used as a basis in making specific, but not general personnel assignments; due process was required in making some removals, such as those for fraud or racism, but not others including one for being "anti-establishment;" and, apart from a regulation requiring a maternity leave very late in the term of a normal pregnancy, such a leave had to be based on an individual medical determination of the employee's ability to continue in her job.

The idiographic approach placed severe strains on accountability in the administrative state. It left public employers in a difficult position because there were no clear, broad, general guidelines on how to resolve questions prior to judicial determination. Not only did each case have to be treated separately on its own merits, but public employers were constantly being second-guessed by the judiciary in their dealing with public employees. The fact that neither the public employer's nor the public employee's constitutional right were self-evident or adequately delineated by court decisions encouraged employees to engage in activities which might be opposed by their employers, tended to deter public employers from taking action against such employees, and led to a substantial increase in the number of such cases reaching the courts. Maintaining accountability through the application of sanctions upon public employees was all but out of the question under these circumstances.

By 1975 a marked shift began to occur. The Nixon bloc had emerged as the dominant force on the Supreme Court. This position was enhanced by the departure of Justice Douglas and the appointment of Justice Stevens. The Court began to make substantial inroads on the idiographic approach by "deconstitutionalizing" this area of the law. "Deconstitutionalization" is best understood as a process of refusing to treat issues previously entertained under the doctrine of substantial interest and idiographic approach or treating them on the basis of statutory law, rather than on constitutional terms. The latter technique has the effect of reducing the Supreme Court's involvement in issues and affords legislators and executives more leeway in regulating public employment. Several examples come to mind. For instance, in *Bishop* v. *Wood* (1976:693), a case dealing with the dismissal of a policeman, the Court reasoned that:

> The federal court is not the appropriate forum in which to review the multitude of personnel decisions that are made daily by public agencies. We must accept the harsh fact that numerous individual mistakes are inevitable in the day-to-day administration of our affairs. The United States Constitution cannot feasibly be construed to require federal judicial review for every such error.

In other cases, the court sought to encourage the treatment of equal opportunity issues under the existing statutes rather than the equal protection clause of the Fourteenth Amendment (*Washington* v. *Davis,* 1976) and to limit the kind of substantive rights involving public employees to be afforded constitutional protection (*Kelly* v. *Johnson,* 1976; *McCarthy* v. *Philadelphia,* 1976). At the same time, however, while the Burger Court is seeking to deconstitutionalize much of the public employment relationship, it nevertheless does not appear that a roll-back to the doctrine of privilege is in the offing. For instance, in *Elrod* v. *Burns* (1976), the Court ruled that patronage dismissals violate the First Amendment rights of most categories of public employees.

From the perspectives of accountability, it is ironic that by the time the Supreme Court began to reduce the constitutional protection afforded public employees, the public sector labor movement had been so successful that little substantive change may be forthcoming.

THE PUBLIC SECTOR LABOR MOVEMENT

Accountability in the administrative state has been further limited by the emergence of public sector labor unions as a political force of major consequence. Despite earlier attempts at union-busting and regulations prohibiting collective bargaining in the public sector, by the 1970s public service labor unions had made tremendous strides. Thus, between 1962 and 1968, public employee organizational membership increased by some 136%, as compared to a 5% increase over the same period in the private sector. By 1970, almost 60% of all federal employees had joined unions involved in collective bargaining with the government. By 1976, more than half of all local employees had done likewise (Shafritz et al., 1978). But numbers only tell part of the story. The evolutionary direction of public sector labor relations has been to follow procedures used in the private sector, including the use of strikes and other job actions. Thus, both in the federal government and the states, the "meet and confer" approach for dealing with public service unions is being replaced by a more fullfledged "collective bargaining" arrangement. The former was a special variant of labor relations found in the public sector through which the rights of management were firmly insisted upon under prevailing notions of "sovereignty." Collective bargaining, on the other hand, opens up many more items for negotiation, including wages, hours, and many matters affecting public policy. It may also include the right to strike, as is true for some public employees in at least six states. Indeed, the strength of public sector labor unions under the collective bargaining approach has been so pronounced that many believe that "codetermination" will be the order of the day in the future.

"Codetermination" is the term applied when public sector labor unions play a major role in the determination of public policy, either on a *de jure* or *de facto* basis. Thus far, codetermination is most pronounced in the realm of public personnel policy. Nigro and Nigro (1976:319) point out that labor relations approaches requiring management to confer with unions before making changes in personnel policy has

> served to make the unions important participants in the personnel function. Even where items may not be negotiated because they are reserved as management rights, the strength of the unions can dissuade management from making certain kinds of personnel decisions.

For example, the federal labor relations program currently enables unions to negotiate such matters as arrangements for employees adversely affected by technological change and the nature of grievance procedures. In addition, postal employees may bargain over wages, hours, and several other terms and conditions of employment.

Codetermination in personnel matters has a marked impact in reducing the power of public managers over rank-and-file public employees. This, of course, reduces accountability by enhancing the general protective characteristic of most public personnel systems. Managers are simply not free to deal with their employees as they think fit. Rather, they must often negotiate what is fit—or at least practicable—with unions. While many think that this may eventually lead to more realistic and better public personnel policy, fewer express such a sanguine view when it comes to codetermination of other kinds of public policy.

Although it is still often marked by a subtlety, public employee unions have made considerable gains in negotiating work conditions which really involve decisions concerning central matters of public policy. For example, it is now a well established practice for teachers to negotiate matters such as class size, procedure for selecting textbooks, school calendar, and preparation periods. As noted above, though unusual, the highly militant and divisive New York City teachers' strike of 1968 was largely over the fundamental matter of local community control of the schools. Similarly, social workers have successfully negotiated maximum case loads; nurses have been able to limit their participation in nonnursing duties in the same way. In one of the most pronounced examples of codetermination of policy,

> after the Attica tragedy, AFSME Council 82 obtained a seven-point agreement with the State Commissioner of Corrections providing for far-reaching reforms, such as employment and training of new correction officers, expenditure of $800,000 for the purchase of new safety and security equipment, and substantial improvements in meals, clothing allowances, and toilet and shower facilities for the inmates. Although this was not in the form of a union contract, it was in writing. [Nigro and Nigro, 1973: 328]

In the future, it should not be surprising to find police negotiating over the nature of patrols, weapons, and the like.

Thus, the labor union movement militates against accountability in several ways. It reinforces civil service protectionism. It affords public employees with real weapons, including job actions and strikes, even though often illegal, with which to fight for what they want. And what they want may have serious repercussions for the nature of public policy. Indeed, a former New York City budget director has said that, "In truth, few changes can be made in work organizations without union approval, and it is seldom forthcoming except at a price. The powerful municipal unions are becoming the most conservative force in city government, providing the strongest pressure against change and improvement, or rational economies" (Nigro and Nigro, 1973: 328). But union leaders, who do the negotiating, are elected by union members, not by the general public. As they gain strength vis-a-vis public managers and elected officials, government automatically becomes less accountable to the public at large. Public policy making increasingly rests in the hands of the leadership of these private organizations that oversee the running of the administrative state. If this formulation seems extreme, one only need look at Western Europe where the tendency is even more pronounced. Thus, in his discussion of public service trade unions in Western Europe, Brian Chapman (1959:296-297) wrote:

> Fifty years ago in most countries public service associations were either friendly societies or banned institutions. Their power now lies in their numerical strength and its potential electoral importance to politicians. This is reinforced by the fact that they are the accepted spokesmen for those employed by the state. Their aim is to protect the interests of the state in so far as those interests coincide with the interests of their members. Indeed, some extremists have held that the body of civil servants is the state. . . . The truth is that people employed in government service are tending to become not only self-governing but also self-employed. All the evidence . . . points in this direction. This drift towards the syndical state machine is one of the unnoticed oddities of the last fifty years.

In sum, it is evident that the labor union movement places major constraints on the possibility of holding public employees accountable to either the general public or to political executives. When this constraint is coupled to the nature of bureaucracy, the character of the public personnel system, and to the constitutional status of public employees in the United States, the barriers to effective accountability become very formidable indeed. Nevertheless, several techniques for obtaining accountability have been employed on a regular basis in the United States.

THE QUEST FOR ACCOUNTABILITY IN THE ADMINISTRATIVE STATE

"STRENGTHENING" THE ELECTED

Given that the general democratic theory of accountability requires that those holding political power be accountable either to the people or to their elected leaders, it is not surprising that several efforts at establishing greater accountability in the administrative state have centered on strengthening the directive and oversight capacities of elected executives and legislatures. These efforts, however, have not satisfactorily returned political power to the hands of the elected, but rather have created new cadres of appointive administrative aides who have emerged as a power center of their own. This process illustrates one of the fundamental facets of the administrative state: it takes bureaucracy to control bureaucracy.

A good example of the bureaucratization of the office of an elected executive is presented by the Executive Office of the President (EOP). It was established in 1939 in an effort to provide the President with the necessary tools to control and hold accountable the sprawling and rapidly growing federal bureaucracy. Indeed, so disorganized and uncontrolled did the bureaucracy appear, that it was often referred to as the "headless fourth branch of government." At first, the EOP appeared to work relatively well, especially given the Rooseveltian penchant for competitive, nonbureaucratic organization. However, by the 1970s it had evolved into a moderately sized bureaucracy. Thus, under President Ford, the White House Office Staff numbered at least 540 and the entire EOP included some 5,000 personnel with an annual budget of about $20 million. Part of this growth has been due to enlarged governmental involvement in the economy, society, and world; part to the desire to maintain a high level of popular support for the President; and part is probably "Parkinsonian" in nature. One telling statistic in this regard is that despite his campaign pledges to reduce the size of the White House Office, President Carter's wife now has half as many staff under her direction as FDR had during World War II.

Although different Presidents organize the White House Office and EOP in various ways, the tendency toward the "institutionalized" Presidency has been overwhleming. Its effects are that the tool intended to establish greater accountability has itself emerged as less than fully accountable. Thus, Cronin (1975:138) observes:

> The presidential establishment had become over the years a powerful inner sanctum of government isolated from the traditional constitutional checks and balances. Little-known, unelected, and unratified aides on occasion negotiate sensitive international commitments by

> means of executive agreements that are free from congressional oversight. With no semblance of public scrutiny other aids wield fiscal authority over billions of dollars.

Perhaps the EOP might none the less be fully accountable to the President, but observer after observer has noted that with its bureaucratization has come diminished Presidential ability to control its actions. Indeed, many would agree with Cronin (1975:139) that "the presidential bureaucracy is becoming a miniaturization not only of important departments and agencies but also of politically important pressure groups and professions." Watergate was but an extreme manifestation of the problem.

In sum, strengthening the EOP has not made the federal bureaucracy substantially more accountable to the President. As Woll and Jones (1975:216-217) summarize it:

> The bureaucracy, sometimes with Congress but often by itself, has frequently been able to resist and ignore Presidential commands. Whether the President is FDR or Richard M. Nixon, bureaucratic frustration of White House policies is a fact of life. Furthermore, the bureaucracy often carries out its own policies which are at times the exact opposite of White House directives. A classic case occurred during the India-Pakistan war in 1971 when the State Department supported India while the White House backed Pakistan.

Thus, "in the end, after thirty years, the effort to help the president in making government work has not succeeded" (Cronin, 1975:159). The same problem exists, though generally to a lesser extent, at the state and local levels.

When we turn our attention to efforts at strengthening the ability of legislatures to engage in oversight of executive branch bureaucracies, we can observe a similar phenomenon. Taking the federal example, congressional efforts at establishing a high level of oversight capability in an attempt to hold the bureaucracy accountable has led to the proliferation of the number and power of its appointive staff personnel. Indeed, only a neophyte would have been startled when Senator Morgan announced on the floor of the Senate in 1976 that "This country is basically run by the legislative staffs of the Members of the Senate and the Members of the House of Representatives. . . . They are the ones who give us advice as to how to vote, and we vote on their recommendations" (Scully, 1977:42). This is especially true in the area of oversight where, according to one knowledgeable observer, over 90% of the work is actually performed by the staff (Newsweek, 1977:20). Similarly, Congressional reliance on the General Accounting Office has been increasing. Indeed, in the minds of some, the fundamental question has become "Who's in charge here?"–the elected or the unelected (Malbin, 1977).

The examples of the Presidential and Congressional efforts to enhance their capability of controlling the federal bureaucracy illustrate some of the fundamental characteristics of the administrative state. These have been distilled by Anthony Downs (1967:262) and put in the form of "laws":

(1) The Law of Imperfect Control: "No one can fully control the behavior of a large organization."
(2) The Law of Counter Control: "The greater the effort made by a sovereign or top-level official to control the behavior of subordinate officials, the greater the efforts made by those subordinates to evade or counteract such control."
(3) The Law of Control Duplication: "Any attempt to control one large organization tends to generate another."

Not surprisingly, many have given up on the possibility of satisfying the need for accountability through efforts to provide executives and legislators with more staff. Instead, they have turned their attention to budgetary techniques and reorganizations in an effort to obtain what might be called a "rational" bureaucracy.

BUDGETS

In 1965 President Johnson introduced program budgeting into the federal bureaucracy on an across-the-board fashion. PPBS (Planning, Programming, and Budgeting System) was presented by many of its advocates as a panacea for most bureaucratic ills, including the lack of accountability. The basic idea was that budgetary decisions should focus on governmental goals and objectives rather than on such traditional items as equipment and personnel. Once decision makers were aware of the price tags on governmental plans or the achievement of policy goals, they would be in a position to make rational choices. In the process, the bureaucracy would fall under the strong control of political executives, the President, and budget analysts, for it would have to state what it was actually doing and how much it cost.

Despite the logic of PPBS, it was quickly run aground on the straits of bureaucratic politics. By 1971 it was formally pronounced dead. Even earlier, however, it had become little more than a "paper program." As Aaron Wildavsky put it: "PPBS has failed everywhere and at all times. Nowhere has PPBS (1) been established and (2) influenced governmental decisions (3) according to its own principles" (1977:255). Its failure lay both in technique and politics. As to technique, "no one could perform the required calculations" (Wildavsky, 1977:245), or at least not enough government budget analysts were properly trained to do so. The political problem was both deeper and more perplexing.

The nature of bureaucratic politics is such, as noted earlier, that specification of program goals in any but the most general terms is often difficult. To state priorities among competing interests is to forego some political support. And to diminish political support is to endanger one's agency. Pluralism begets compromise, tacit agreement, and unclarity. Again Wildavsky (1977:255) is helpful:

> Program budgeting is like the simultaneous equation of governmental intervention in society. If one can state objectives precisely, find quantitative measures for them, specify alternative ways of achieving them by different inputs of resources, and rank them according to desirability, one has solved the social problems for the period. One has only to bring the program budget up to date each year. Is it surprising that program budgeting does not perform this sort of miracle? Even a modified version–in which all activities are placed in programs that contribute to common objectives, but objectives are not ranked in order of priority–is far beyond anyone's capacity.

No wonder that PPBS, despite its early billing, turned out to be unsatisfactory as a mechanism for controlling the federal bureaucracy and holding it accountable. What is remarkable, however, is that those who remain wedded to the traditional approaches for establishing accountability continue to fail to grasp the fundamental nature of the administrative state.

Thus, after a brief hiatus, a new budgetary technique was in the offing. The Carter Administration has pledged to make Zero Base Budgeting (ZBB) a reality in the federal government. Like PPBS, the major political purpose of ZBB is to allow elected and politically appointed officials to make better, more rational budget decisions (Pyhrr, 1973). This in turn enhances both political control and accountability of the bureaucracy. However, ZBB requires some of the same operations as PPBS–namely, the specification of objectives and the evaluation of programs–and consequently is likely to meet a similar fate. There is no need to belabor the point as it has been made by Peter Pyhrr, the creator of the ZBB concept. He has called Carter's plan "all screwed up" and "absolute folly," largely because it is being introduced on an across-the-board basis, in a too limited time frame, and without adequate preparation (*Houston Post,* 1977:14A). While ZBB may be useful in some contexts, it nevertheless runs against the grain of bureaucratic politics and is not likely to have much of an impact on decisions affecting the most fundamental political issues of the day.

REORGANIZATION

Reorganization schemes tend to share with PPBS and ZBB the fundamental premise that public bureaucracies can be transformed into rational organizations operating with maximum efficiency and effectiveness. In theory, should this occur, elective and appointive political officials would have little difficulty establishing policy objectives and holding the bureaucrats accountable for their implementation. As Harold Seidman (1970:29) points out, this approach holds that "Straight lines of authority and accountability cannot be established in a nonhierarchical system." Consequently:

> Orthodox theory is preoccupied with the anatomy of Government organization and concerned primarily with arrangements to assure that (1) each function is assigned to its appropriate niche within the Government structure; (2) component parts of the executive branch are properly related and articulated; and (3) authorities and responsibilities are clearly assigned. [Seidman, 1970:5]

This approach is fraught with difficulty in the real world of bureaucratic politics.

One major problem with using reorganization in an effort to gain accountability is that bureau shuffling changes neither the fundamental nature of bureaucratic organization nor necessarily brings the polity closer to a resolution of its most pressing problems. To a large extent reorganizations involve moving bureaus from one department or agency to another and consolidating bureaus or agencies into departments. Although overlaps, redundancies, and the number of units to be supervised by political executives may be reduced in this fashion, there is still little guarantee that the latter will be able to control their organizations or hold individual bureaucrats accountable. Indeed, reorganization is used so frequently that it may actually have a detrimental effect on bureaucratic performance. Thus, there is little reason to believe that the reorganization of HUD some 20 times in the past decade has truly served the needs of providing Americans with better housing and better urban environments. Similarly, the constant reorganization of the Federal Equal Employment Opportunity program from its inception in the 1940s until 1965 contributed to its ineffectuality (Rosenbloom, 1977a). Where there are no answers to social and economic problems, reorganization will not supply them. Seidman (1970:3-4) sums it up nicely: "The myth persists that we can resolve deepseated and intractable issues of substance by reorganization."

Orthodox organization theory is on even weaker ground when it assumes that the chief goals to be obtained from public bureaucracies are efficiency and economy. The fact of the matter is that public bureaucracy is political to the core, and efficiency and effectiveness are not high-ranking political values. Power, responsiveness, and representation, by contrast, are central items on any political agenda, and it is at these values that a meaningful theory of public organization must be aimed. As Justice Brandeis aptly put it, "The doctrine of separation of powers was adopted by the Constitution in 1787, not to promote efficiency but to preclude the exercise of arbitrary power" (Seidman, 1970:27). In Seidman's (1970:13) words: "Established organization doctrine, with its emphasis on structural mechanics, manifests incomplete understanding of our constitutional system, institutional behavior, and the tactical and strategic uses of organization structure as an instrument of politics, position, and power." Hence, in reality, reorganization is a political tool, generally aimed at assuring a desired allocation of power and generating a desired amount of support. It is not necessarily or even frequently genuinely intended to make the bureaucracy more rational or more accountable. Given the pluralistic organization of the federal bureaucracy, this is all but inevitable. In sum, there is no reason to believe that the myriad reorganizations of the past four decades have even made a dent in the problem of assuring accountability in the administrative state.

BEYOND ACCOUNTABILITY?

It is evident that given the present organizational, legal, constitutional, and political features of the modern American administrative state, public bureaucrats can only be held accountable to elected and politically appointive officials to a very limited degree. Whatever the proponents of traditional democratic theory or political authorities find comforting to believe in this regard, the undisputable reality is that the administrative state places a great deal of unchecked political power in the hands of relatively permanent bureaucratic functionaries and organizations. Nor have the traditional approaches for dealing with this political fact of life been more than marginally successful. In short, in the administrative state, power flows to the unelected, the vast majority of whom are appointed without regard to politics and are well insulated from political control. Since obtaining a satisfactory degree of accountability under these circumstances appears impossible, many have questioned whether it is not desirable to go beyond these traditional concerns and attempt to capitalize on the potential political virtues of the administrative state in some other fashion.

Two somewhat opposing lines of thought have emerged in this regard. First, some, along with Herbert Storing (1964:154-155) have maintained that the political insulation of the civil service is one of its chief political advantages:

> The special kind of practical wisdom that characterizes the civil servant points to a more fundamental political function of the bureaucracy, namely to bring to bear on public policy its distinctive view of the common good or its way of looking at questions about the common good. . . .Like judges, civil servants have a special responsibility to preserve the rule of law.
>
> Civil servants also bear a similarity to judges in their possession of what is, for most practical purposes, permanent tenure in office. Of course, like judges, they are influenced by the election returns—and it would be dangerous if they were not; but they have a degree of insulation from shifting political breezes.
>
> At its best, the civil service is a kind of democratic approximation to an hereditary aristocracy whose members are conscious of representing an institution of government which extends into the past and into the future beyond the life of any individual member. In our mobile democracy, the civil service is one of the few institutions we have for bringing the accumulated wisdom of the past to bear upon political decisions.

In short, if the civil service cannot be held accountable in any strict or satisfactory sense, the society might nevertheless benefit from the agglomeration of bureaucratic power by making use of the civil service's "institutional qualities which give it a title to share with elected officials in rule" (Storing, 1964:156). Independence, in this view, is not necessarily unhealthy.

A second line of thought for moving beyond concern with traditional notions of accountability stresses the possibility that a bureaucracy can be a representative and responsive political institution. Here, the main interest is in organizing and staffing public bureaucracies in such a fashion that they represent and reflect the major political groupings of the society at large. Organizationally, for instance, some have maintained that the myriad bureaus of United States federal bureaucracy are highly representative of most major economic and social groupings found within the political community. In terms of personnel, proponents of "representative bureaucracy" argue that "Who writes the [administrative] directive—his or her style, values, concept of role—is as significant as who gets to be president, congressman, senator, member of parliament, or cabinet minister. The notion of representative bureaucracy is that broad social groups should have spokesmen and officeholders in administrative as well as political positions" (Krislov, 1974:7). Indeed, many support a more extreme formulation:

> As it operates in the civil service, the recruitment process brings into federal employment and positions of national power, persons whose previous affiliations, training, and background cause them to conceive of themselves as representing constituencies that are relatively uninflu-

> ential in Congress. These constituencies, like that of the presidency, are in the aggregate numerically very large; and in speaking for them as self-appointed, or frequently actually appointed, representatives, the bureaucrats fill in the deficiencies of the process of representation in the legislature. [Long, 1962:70]

From the perspectives of democratic theory, it stands to reason that if a bureaucracy were a representative institution, and especially if it were the most representative element in the political system, there would be little desirability in holding it accountable to the less representative branches of government. However, the extent to which public bureaucracies can be representative in this sense remains hotly contested (Dresang, 1974; Mosher, 1968; Rosenbloom, 1977a; Nachmias and Rosenbloom, 1978). Consequently, many believe it is too soon to give up entirely on traditional concerns with accountability.

Finally, a word should be said about the future, although here, the issues are even less clear. In the view of some, the *bureaucratic* administrative state is a legacy of the past which will be outgrown in the years to come. Those who believe we will eventually move "beyond bureaucracy" (Bennis, 1969) argue that administrative organizations will become more democratic in the future. Should this occur, their internal checks and balances will multiply, and presumably contribute to the formulation of better public policy, greater efficiency, and more effectiveness in the process. In the view of others, however, this amounts to little more than wishful thinking (Golembiewski, 1977).

In conclusion, it is evident that the administrative state presents an overwhelming challenge to traditional concerns with accountability. Not only have traditional mechanisms failed to satisfactorily check and contain bureaucratic power, but many now believe that the key to successfully integrating bureaucratic power with democracy lies elsewhere. Until a resolution or new formulation of this fundamental problem is developed, we will continue to be vexed by Mosher's (1968:4) question: "How can we be assured that a highly differentiated body of public employees will act in the interests of all the people, will be an instrument of all the people?"

NOTE

1. Parts of this section are drawn from Rosenbloom (1977b).

REFERENCES

Arnett v. Kennedy (1974). 416 U.S. 134.

Bailey v. Richardson (1950). 182 F2d 47; 341 U.S. 918.

BARKER, E. (1966). The development of public services in Western Europe. Hamden, Conn.: Archon.

BARTLETT, J., and JONES, D. (1974). "Managing a cabinet agency: problems of performance at commerce." Public Administration Review, 34(January/February):62-70.

BENNIS, W. (1969). "Beyond bureaucracy." Pp. 1-8 in A. Etzioni (ed.) Readings on modern organizations. Englewood Cliffs, N.J.: Prentice-Hall.

Birnbaum v. Trussell (1966). 371 F2d 672.

Bishop v. Wood (1976). 48 L. Ed2d 684.

BLAU, P., and MEYER, M. (1971). Bureaucracy in modern society. New York: Random House.

Board of Regents v. Roth (1972). 408 U.S. 564.

CHAPMAN, B. (1959). The profession of government. London: Unwin University Books.

CRONIN, T. (1975). The state of the presidency. Boston: Little, Brown.

DOWNS, A. (1967). Inside bureaucracy. Boston: Little, Brown.

DRESANG, D. (1974). "Ethnic politics, representative bureaucracy and development administration: the Zambian case." American Political Science Review, 68(December):1605-17.

Elrod v. Burns (1976). 49 L. Ed2d 547.

GAWTHROP, L. (1969). Bureaucratic behavior in the executive branch. New York: Free Press.

Gilmore v. James (1967). 274 F. Supp. 75.

GOLEMBIEWSKI, R. (1977). "A critique of 'democratic administration' and its supporting ideation." American Political Science Review, 71(December):1488-1507.

HALL, R. (1972). Organizations. Englewood Cliffs, N.J.: Prentice-Hall.

HERSHEY, C. (1973). Protest in the public service. Lexington, Mass.: Lexington Books.

Houston Post (1977). April 8, p. 14a.

HUMMEL, R. (1977). The bureaucratic experience. New York: St. Martin's Press.

Kelly v. Johnson (1976). 47 L. Ed2d 708.

KRAMER, F. (1975). "Policy analysis as an ideology." Public Administration Review, 35(September/October):509-517.

KRISLOV, S. (1974). Representative bureaucracy. Englewood Cliffs, N.J.: Prentice-Hall.

LONG, N. (1965). "Power and administration." Pp. 14-23 in F. Rourke (ed.), Bureaucratic power in national politics. Boston: Little, Brown.

--- (1962). The polity. Chicago: Rand McNally.

McCarthy v. Philadelphia Civil Service Commission (1976). 47 L. Ed2d 366.

MALBIN, M. (1977). "Congressional committee staffs: who's in charge here?" The Public Interest, 47(Spring):16-40.

MOREHOUSE, T. (1972). "Program evaluation: social research versus public policy." Public Administration Review, 32(November/December): 868-874.

MOSHER, F. (1968). Democracy and the public service. New York: Oxford University Press.

NACHMIAS, D. (1978). Policy analysis and evaluation. New York: St. Martin's Press.

--- and ROSENBLOOM, D. (1978). "Bureaucracy and ethnicity." American Journal of Sociology, 84(January).

NATHAN, R. (1975). The administrative presidency. New York: Wiley.

Newsweek (1977). January 17, p. 20.

New York Times (1974). July 20, p. 14.

NIGRO, F., and NIGRO, L. (1976). The new public personnel administration. Itasca, Ill.: F. E. Peacock.

--- (1973). Modern public administration. New York: Harper & Row.

PHYRR, P. (1973). Zero base budgeting. New York: Wiley.

POSEY, R. (1968). "The new militancy of public employees." Public Administration Review, 28(March/April):111-117.

ROSENBLOOM, D. H. (1977a). Federal equal employment opportunity. New York: Praeger Special Studies.

--- (1977b). "The public employee in court: implications for urban government." Pp. 57-82 in C. Levine (ed.), Managing human resources. Beverly Hills: Sage Publications.

--- (1971). Federal service and the constitution. Ithaca, N.Y.: Cornell University Press.

--- and GILLE, J. A. (1975). "The current constitutional approach to public employment." University of Kansas Law Review, 23(2):249-275.

SAYRE, W. (1965). The federal government service. Englewood Cliffs, N.J.: Prentice-Hall.

SCULLY, M. (1977). "Reflections of a Senate aide." The Public Interest, 47(Spring):41-48.

SEIDMAN, H. (1970). Politics, position, and power. New York: Oxford University Press.

SHAFRITZ, J., BALK, W., HYDE, A., and ROSENBLOOM, D. H. (1978). Personnel management in government. New York: Marcel Dekker.

SMITH, B., and HAGUE, D. (1971). The dilemma of accountability in modern government. New York: St. Martin's Press.

STORING, H. (1964). "Political parties and the bureaucracy." Pp. 137-158 in R. Goldwin (ed.), Political parties, U.S.A. Chicago: Rand McNally.

Sugarman v. Dougall (1973). 413 U.S. 634.

THOMPSON, V. (1961). Modern organization. New York: Knopf.

U.S. Civil Service Commission (1976). "A self-inquiry into the merit system." Washington, D.C.: Author.

U.S. Government Organization Manual (1972/73). Washington, D.C.: Government Printing Office.

Washington Post (1970). October 10, p. B1.

Washington v. Davis (1976). 48 L. Ed2d 597.

WEBER, M. (1958). From Max Weber: essays in sociology (H. H. Gerth and C. W. Mills, trans. and eds.). New York: Oxford University Press.

WILDAVSKY, A. (1977). "PPBS." Pp. 244-256 in A. Altshuler and N. Thomas (eds.), The politics of the federal bureaucracy. New York: Harper & Row.

WOLL, P., and JONES, R. (1975). "Bureaucratic defense in depth." Pp. 216-224 in R. Pynn (ed.), Watergate and the American political process. New York: Praeger.

5

Performance Indicators in the Court System

JAMES L. GIBSON

□ WELL OVER A CENTURY AGO Alexis de Tocqueville observed that "Scarcely any political question arises in the United States that is not resolved, sooner or later, into a judicial question." It is difficult to imagine a more appropriate way of describing the relationship between courts and public policy in modern day America. Many, if not most, of the pressing political issues of the last few decades have been the object of judicial involvement and pronouncement. Policy outputs on school desegregation, reapportionment, law and order, civil liberties, civil rights, to name only a few, have been influenced at least as much by judges as by legislators. Economic and foreign policies are perhaps the only policy areas in which courts are not coequal policy makers (Horowitz, 1977).

The policy-making activity of the courts has not been subjected, however, to the intense public scrutiny nor mechanisms of accountability commonly found in representative democracies. This is in part a function of the heavy veil of symbols which is attached to judicial decisions. Despite the incisive observations of de Tocqueville, not all segments of society recognize the vast policy-making power of the judiciary. Because the activities of courts are perceived by mass publics as primarily mechanical, discretionless, and technical, rather than as political and value-laden (Casey, 1974; and this perception may even be shared by lawyers: Beiser, 1972), judging is frequently perceived as simply laying "the article of the Constitution which is invoked beside the statute which is challenged and . . . deciding whether the latter

AUTHOR'S NOTE: *This is a revised version of a paper delivered at UWM Urban Research Center Conference: "Public Agency Accountability in an Urban Society," (Milwaukee, Wisconsin) April 3-5, 1977.*

squares with the former." This perception reduces accountability in policy-making to the question of technical competence. In the absence of any awareness of the policy consequences of decisions, judicial policy making becomes insulated from political control. Many judicial actors are keenly aware of the necessity of maintaining the appearance of legal technicians precisely so that accountability does not become an issue.[1]

The lack of understanding of judicial policy making may stem from the lack of research which conceptualizes and measures the policy outputs (in contrast to the administrative efficiency) of courts, especially lower courts. Few efforts have been made to develop performance indicators for courts (but for an exception see Wildhorn et al., 1976), so it is not surprising that courts have not been evaluated as policy-making institutions.[2] Thus, the first step in any discussion of courts and accountability must be the conceptualization and measurement of the policy outputs of courts.

This research represents an effort to conceptualize (from a functionalist position) and measure the policy-making activities of criminal trial courts. Indicators of court performance on several policy dimensions are developed for 25 county courts in Iowa. After describing court outputs, the research offers some tentative findings regarding (a) factors accounting for variations in court policy making, and (b) the impact of court policies on the jurisdiction in which the court operates. The research concludes with some observations on making courts politically accountable for the policies they make.

TRIAL COURT POLICY MAKING

In order to understand policy making at the trial level, it is useful to make some definitions explicit. Policy making is a process in which decisions, authoritatively allocating values for a significant proportion of society, are made under conditions of broad, formal and informal decision-making discretion. Policy making can be distinguished from policy implementation primarily by the degree of discretion in decision making. There is no doubt, for instance, that administrative hearings (e.g., social security, welfare, and so on) authoritatively allocate values, but there are fairly narrow limits on the discretion attached to these decisions. Policy making must include both elements.[3]

Trial courts undoubtedly authoritatively allocate values in their decisions. The conceptual question concerning trial court policy making is thus the degree to which trial court judges exercise discretion instead of implementing public policy made by other institutions.[4] That is, what formal and informal constraints exist to direct trial court decision making? Very few indeed.

Criminal courts are not simply "enforcing" the "law" as determined by the legislature: just as with appellate courts, the power to "interpret" the law is so broad and so filled with discretion that the criminal courts become, de facto, an active maker of public policy. For instance, bail must only be "reasonable" to satisfy the law. Sentencing options are so broad that in nearly *every single crime* (no matter how severe) in the United States the sentencing judge can select immediate probation for the offender. Indeed, the options are so broad that some judges have complained bitterly (Frankel, 1973). Certainly with regard to formal courtroom proceedings, the judge possesses great power in both legal and factual questions, and appeals, as a percentage of all cases, are too infrequent to be an effective means of controlling trial judge discretion.[5] Without legal limits on the decisional options available, judicial decision making should be considered as an instance of policy making, rather than policy implementation.

But what type of policies do criminal courts make? In order to consider this question it may be useful to placc the policy activity of criminal courts in a functional perspective. By specifying the functions of these courts, a basis for comparing the outputs of the court systems is provided. Variation in outputs can then be taken as evidence (with proper controls, of course) of policy-making activity on the part of the court system. The question becomes: what functions do criminal courts perform and do different systems perform the functions differently? (These are followed, of course, by the questions of why the systems vary and with what consequences.)

The functions of criminal trial courts are to identify those who can be legitimately sanctioned, and to apply the sanctions. This implies three different dimensions of the policy outputs of trial courts: (1) the identification of defendants and the determination of their guilt or innocence; (2) the sanctioning of the convicted; and (3) the legitimacy, or procedural properness, with which the processing of defendants takes place. The identification and determination of guilt can be measured primarily in terms of "effectiveness"—the extent to which those committing crimes are apprehended and processed by the criminal justice system. Sanctions can be measured primarily in terms of "punitiveness," and legitimacy refers to the extent to which the first two activities are performed with procedural due process. These three dimensions will be used to characterize the policy outputs of the criminal trial courts in Iowa.

DATA

This research uses as its data base the Iowa district court system. The district courts are the state trial courts of general and original jurisdiction, and are typical of many state trial courts. All criminal charges are either felonies or indictable misdemeanors. In noncriminal matters, the district courts try most major civil actions. The system is organized into 99 county courts. There are eight judicial districts in Iowa, with each district composed of a number of counties. This research focuses upon the three southeastern districts of the system. While the districts were selected primarily on ease of access criteria, they are representative, at least on several aggregate indicators, of the state trial court system. Twenty-five of the 99 counties and 24 of the 83 judges are included in these three districts.

The universe of criminal cases analyzed includes all cases initiated during 1972 and 1973 and concluded by the end of 1974. The data were collected from criminal case files stored at each of the county courthouses. In 22 of the counties no sampling was necessary, so the universe of cases was selected. In the remaining three, the three most populous counties in the districts, random sampling was employed. This procedure resulted in selecting 5,350 cases. In 2,715 cases, 50.7% of the total, convictions resulted.

EFFECTIVENESS

The "effectiveness" of the criminal justice system is the success at "solving" crimes. It is the degree to which those committing crimes are identified, apprehended, and convicted. It is the capability of the system for dealing with criminal behavior. Except at the crudest level, effectiveness represents insurmountable measurement problems. However, it is possible to ascertain the rate at which the court system initiates formal action to determine guilt or innocence. That is, given certain assumptions, the percentage of offenses committed in which convictions resulted can be measured. Several assumptions, some which may be problematic, must first be made. Nevertheless, even a crude indicator of the effectiveness of the court system's response to criminal behavior within its jurisdiction may be useful.

Effectiveness is affected by the various means by which cases are diverted out of the criminal justice system. Figure 1 illustrates, in simplified form, the various steps in "solving" crimes. At each step a number of crimes are diverted out of the system by a decision of one or more of the principal actors. Except at the reporting stage, the decisions are made primarily by criminal justice officials. These decisions reflect both competence (e.g., the competence of the police at apprehending criminals) and policy (e.g., the decision not to prosecute certain offenses). The effectivencss of the criminal justice system (which is obviously affected by competence and policy) can thus be seen in the percentage of reported crimes resulting in apprehension, charges, and convictions.

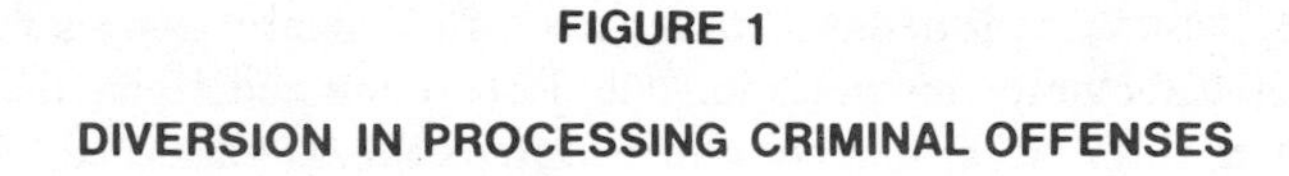

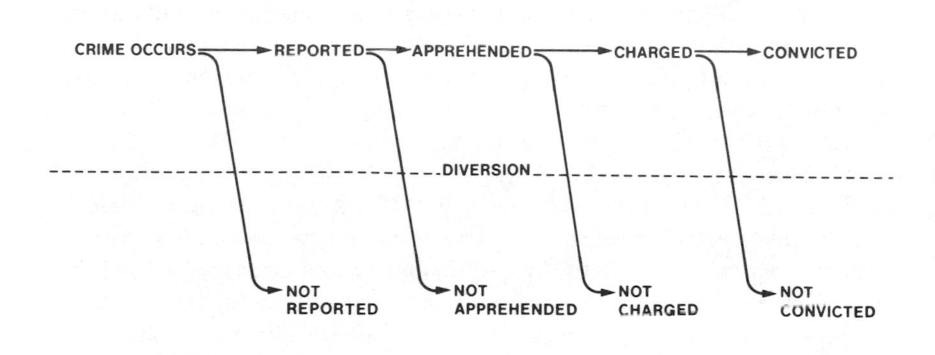

In order to arrive at a measure of effectiveness, an indicator of the total amount of reported crime in the environment of the court must first be available. The two alternatives for such a measure are (1) the FBI's Uniform Crime Reports (UCR), and (2) victimization surveys. Victimization surveys are to some extent a more reliable measure of crime than UCR figures (although serious problems still exist with victimization measures) because a large proportion of the crimes committed in the United States are never reported to the police and hence are not included in UCR counts. The surveys are, however, expensive and have only been conducted in a few areas. Furthermore, as Skogan (1974) has argued, the degree of nonreporting may be relatively constant across jurisdictions and, at least some crimes, such as murder and motor vehicle theft, are reported at a very high rate. Finally, the decision to report a crime is not a decision primarily controlled by criminal justice officials, and thus reported crime should serve as the standard for effectiveness. Despite their shortcomings, UCR data are available and therefore must serve as the indicator of the amount of crime in each of the counties.

Effectiveness can be measured therefore in terms of (a) the percentage of offenses reported to the police in which formal charges were filed and (b) the percentage in which convictions resulted. This requires, however, an assumption about the time frame for the crime and court action. Because no offender-based transaction statistics (cf. Pope, 1975) exist to link FBI and court records, it must be assumed that (a) crimes committed in a particular time period are prosecuted within that time period, or (b) that the number of crimes held over from the preceding time period is equal to the number delayed until the next time period. We know, for example, that 23 murders were reported to the police in these counties in 1972 and 1973, and that 16 charges of murder were filed, resulting in 13 convictions during the same period; but we do not know for certain (because of time lags in apprehension, trial, and so on) that the reported and charged murders represent the same incidents. In order to construct an effectiveness measure, it must therefore be assumed that most crimes are prosecuted in the year in which the crime was committed (if at all) and that, for the remaining offenses there is a balance between crimes occurring before the time period under study and charges occurring after the period under study.[6]

One further difficulty exists: since the UCR categories are more general than the charges filed in court, a single UCR crime type may result in a wide variety of charges. Thus, a burglary in Iowa may result in charges for burglary with or without aggravation; possession of burglary tools; breaking and entering; entering in the nighttime; and attempting to break and enter (the

TABLE 1
COURT SYSTEM EFFECTIVENESS

	Murder[a]			*Assault*[b]		
County	Reported	Charges Filed	Convictions	Reported	Charges Filed	Convictions
1	1	0	–	8	3	0
2	0	–	–	6	4	1
3	1	0	–	6	1	0
4	5	2	2	36	16	10
5	0	1	1	4	6	2
6	7	3	2	28	8	2
7	1	1	1	11	2	2
8	0	–	–	4	0	–
9	0	–	–	2	2	1
10	2	2	2	25	1	0
11	0	–	–	2	0	–
12	0	–	–	0	3	2
13	2	2	1	32	22	10
14	0	–	–	31	3	1
15	1	0	0	41	2	1
16	0	–	–	9	0	–
17	0	–	–	66	19	9
18	0	–	–	12	2	1
19	3	3	3	13	10	6
20	0	–	--	7	2	2
21	0	2	1	21	12	6
22	0	–	0	6	3	0
Total	23	16	13	370	120	56

full set of UCR–court equivalences is shown in Table 1). Further, the FBI counts multiple offense crimes as a single crime, with the most serious offense receiving credit for the incident. Prosecutor charges, however, may reflect any of the component crimes. This problem is somewhat ameliorated by aggregation of the seven specific offenses to an index of serious crime, so that the precise charge, so long as it is one of the seven index crimes, becomes somewhat less crucial.

Despite these somewhat limiting assumptions, an effort was made to measure the rate at which formal charges were filed and the conviction rate, both as percentages of the total crimes reported in the county. These figures are shown in Table 1. The data reveal a few instances in which the assumptions have been clearly violated: the number of charges filed exceeds the number of reported crimes. This is true for counties 5 and 21 for murder cases; counties 5 and 12 for assault charges; counties 5, 9, 14, 17, and 19 for rape charges; and counties 11, 14, and 17 for robbery. In every instance of obvious error, a small number of offenses is involved. (This is not to say that

TABLE 1 (Continued)

		Rape[c]			*Robbery*[d]	
County	Reported	Charges Filed	Convictions	Reported	Charges Filed	Convictions
1	8	2	1	3	1	0
2	3	2	1	1	0	0
3	4	2	2	1	0	0
4	17	3	1	20	5	3
5	0	1	0	0	–	0
6	9	3	2	10	3	1
7	1	1	1	2	0	0
8	1	0	–	0	–	0
9	0	1	0	1	1	1
10	3	3	2	7	1	1
11	0	–	–	0	1	1
12	1	0	–	0	–	0
13	1	1	0	8	2	2
14	3	4	2	0	4	4
15	3	2	1	10	4	1
16	1	1	0	2	0	0
17	3	8	2	11	12	9
18	2	1	1	4	0	0
19	0	2	1	1	1	1
20	1	0	–	2	1	0
21	5	4	4	21	7	4
22	2	0	–	1	0	0
Total	68	41	21	105	43	28

TABLE 1 (Continued)

County	Burglary[e] Reported	Burglary[e] Charges Filed	Burglary[e] Convictions	Larceny[f] Reported	Larceny[f] Charges Filed	Larceny[f] Convictions
1	188	6	5	435	3	2
2	159	12	5	173	4	1
3	108	1	0	172	3	1
4	538	18	14	2333	10	9
5	55	6	4	101	4	2
6	347	18	5	975	18	8
7	158	1	1	291	7	7
8	78	1	1	143	7	7
9	82	11	11	256	3	1
10	213	8	5	461	5	3
11	77	5	5	300	5	2
12	75	0	–	169	7	5
13	601	31	16	1263	28	13
14	183	5	3	193	4	3
15	360	3	2	607	16	13
16	102	4	2	214	5	4
17	164	17	12	1134	13	10
18	135	6	3	289	2	2
19	198	13	9	150	13	5
20	62	7	6	90	2	1
21	508	13	2	606	17	11
22	145	1	1	288	2	2
Total	4986	187	112	10,643	178	112

error does not exist in the other instances: it may exist, but it is not obvious.) The error stems from violations of the assumptions.

These data nevertheless portray a quite dismal image of criminal justice effectiveness. Only murders and rapes result in charges in more than one-half of the offenses, and only murderers face a probability of conviction greater than .5. The rate for burglary, larceny, and auto theft are extremely low, with only 2.8% of the offenses committed resulting in charges or convictions. Overall, the criminal justice system is not very effective in its response to crime.

On the whole, however, these figures probably underestimate criminal justice effectiveness. That is, effectiveness is probably no worse than these figures indicate. For instance, for a small number of cases the arrest year could not be ascertained so the number of charges filed may be somewhat underestimated. While the convictions are for any offense, not just the offense charged, the charging and conviction measures do not reflect charges

TABLE 1 (Continued

County	Reported	*Auto Theft*[g] Charges Filed	Convictions	Reported	*Total Index Crimes* Charges Filed	Convictions
1	32	2	1	675	17	9
2	19	0	–	361	22	8
3	23	5	2	316	12	5
4	191	4	3	3140	57	42
5	16	2	0	176	20	9
6	94	13	3	1470	66	23
7	25	3	1	489	15	13
8	5	0	–	227	8	8
9	9	3	2	350	21	16
10	32	3	1	743	23	14
11	13	1	1	392	12	9
12	1	1	0	246	11	7
13	150	12	9	2057	98	51
14	10	0	–	420	20	13
15	51	5	4	1073	32	22
16	16	0	–	344	10	6
17	43	2	2	1877	71	44
18	18	3	3	456	14	10
19	23	3	1	388	45	26
20	4	1	1	166	13	10
21	206	10	6	1367	65	34
22	24	3	3	467	9	6
Total	1005	76	43	17,200	661	385

[a]The following charges have been classified as murder charges: first degree murder, second degree murder, and manslaughter.

[b]The following charges have been classified as aggravated assault charges: assault with intent to commit murder; poisoning food or drink with intent to kill; mayham; assault and battery; pointing a gun at another; assault while masked; assault with intent to commit a felony; assault with intent to commit great bodily harm; and assault with intent to maim.

[c]The following charges have been classified as rape charges: rape; and assault with intent to commit rape.

[d]The following charges have been classified as robbery charges: assault with intent to commit robbery; robbery with aggravation; and robbery without aggravation.

[e]The following charges have been classified as burglary charges: burglary without aggravation; burglary with aggravation; possession of burglary tools; breaking and entering; entering in the nighttime; and attempting to break and enter.

[f]The following charges have been classified as larceny charges: larceny; larceny in the nighttime; larceny in the daytime; attempted larceny; larceny from a person; larceny of gas; larceny of domestic animals; appropriating found property; larceny-common thief; shoplifting; and larceny from a parking meter. All larceny charges involve property valued at $20 or more.

[g]The following charges have been classified as motor vehicle theft charges: operating a motor vehicle without the consent of the owner; possession of a stolen vehicle: larceny of a motor vehicle; and altering or changing a vehicle identification number.

filed for minor misdemeanors in lower courts. Thus, to some degree the criminal justice is probably more effective than these estimates suggest.

Beyond the question of the amount of error in the estimates lies an interpretation problem: what the figures mean and what factors account for the low rates is not altogether clear. The low rates may reflect the ineffectiveness of segments of the criminal justice system other than the courts, and particularly the police. If the police fail to make arrests, then obviously charges cannot be filed. The "clearance" rate[7] further restricts the pool of offenders who might be charged. If Iowa is representative of the rest of the nation, then the clearance rates for the seven index crimes are shown in Table 2. By comparison the percentage charged, and the percentage convicted are also shown. It seems that only a small proportion of those available to be charged are indeed charged, and that even fewer of those charged are convicted. Thus, the law enforcement system obviously affects these charging and convictions rates, but there is probably also an independent impact of the actions of the prosecutor and the court system.

The low rates may also reflect conscious policy decisions by judges, prosecutors, and police not to invoke the formal machinery of the law against offenders. Instead, informal sanctions may be deemed sufficiently severe to accomplish the purposes of the criminal justice system. For instance, arrest alone may be sufficient to deter future criminal behavior where the crime is minor. If the charging and conviction decisions do reflect policy, then the percentages should be related and in fact the correlation exceeds .8.

There is also considerable variation among the court systems in effectiveness. In terms of charging percentages, the systems vary from a rate of 1.8% to 11.6% for the index of seven crimes. Conviction rates vary from 1% to 6.7%. Charging rates for five of the seven crimes vary from 0% to 100%, and

TABLE 2
CLEARANCE RATES AND COURT EFFECTIVENESS

Crime	Percent Cleared[a]	Percent Charged[b]	Percent Convicted[b]
Murder	75-80	70	57
Rape	50-55	60	31
Aggravated Assault	60-67	32	15
Robbery	25-30	41	27
Burglary	15-20	4	2
Larceny	15-20	2	1
Motor Vehicle Theft	10-15	8	4

[a] FBI national estimates.
[b] From Table 1.

convictions rates also exhibit extreme variability across counties. The variation may be a function of differences in public policy on crime.

In summary, the effectiveness of Iowa court systems does not appear to be overwhelming. The rate at which offenses result in charges and/or convictions for most crimes is extremely low, even when discounted for the impact of clearance rates. This may be a reflection of policy decisions to avoid formal adjudication procedures. Finally, considerable variability among the various systems exists.

PUNITIVENESS

The most salient dimension of the outputs of courts is the sanction imposed on convicted criminals. A major function of criminal courts is to apply sanctions—to use the coercive power of the state. By law, however, much discretion exists in the use of this power so it is likely that systems will differ in the severity of the sanctions they apply. Thus, punitiveness is the second dimension of the outputs of criminal courts which must be measured.

Punitiveness has been measured traditionally by the severity of criminal sentences. While this is indeed an important dimension of court sanctions, there are several other decisions made by the system which may also reflect punitive policy orientation. Bail decisions, decisions in pre-trial hearings (e.g., search and seizure, preliminary hearings, and so forth), and decisions at trial all represent ways in which defendants can be sanctioned. If there is a general punitiveness policy orientation, then the various decisions should be related, despite the fact that substantively they involve quite different issues.

This hypothesis can be tested using measures of some but not all of the decisions identified above. Specifically, four variables are used:

(1) the average sentence severity, corrected for the seriousness of the crime;[8]

(2) the average amount of bail, corrected for the seriousness of the crime;[9]

(3) the percentage of defendants unable to make bail; and

(4) the average time between arrest and bail.

If a common dimension exists, then each of these should be highly related to the other (and presumably to the other decisions already identified but not measured). One way in which to test the hypothesis is to factor analyze the variables to see if they all load on a single common dimension. Table 3 shows the results of the principal components factor analysis.

TABLE 3
FACTOR ANALYSIS OF PUNITIVENESS INDICATORS

Indicator	Factor Loadings[a]
Average Sentence Severity	.66
Average Amount of Bail	.65
Percent Unable to Make Bail	.79
Average Time between Arrest and Bail	.74
Eigenvalue	2.03
% of Variance Explained	50.7

[a]Since only a single factor had an eigenvalue exceeding 1 no rotation was necessary.
N = 25.

The analysis supports the hypothesis: a single common dimension, punitiveness, does seem to underlie the various decisions. Sentence severity is only one of the many interrelated aspects of punitiveness. It should be noted additionally that it is not the seriousness of the offense which accounts for the relationships because the variance attributable to charge seriousness has been purged.

Very considerable variation across systems exists in the factors scores from the punitiveness dimension and the distribution is skewed toward nonpunitiveness. Two thirds of the systems are below the mean, while only five systems are greater than one standard deviation above the mean. Apparently, these systems differ enough in their policies on punitiveness that similar offenses are sanctioned dissimilarly.

DUE PROCESS

Unlike legislative institutions, court outputs are not limited to substantive decisions but include a procedural dimension as well. It is not simply who is sanctioned how severely, or how effective courts are performing their functions, but, due to the tremendous implications of the use of judicial power, *how* the decisions are reached is also vital to understanding outputs of courts. Obviously, this raises the question of due process. Due process requires "formal, adjudicative, adversary fact-finding processes in which the factual case against the accused is publicly heard by an impartial tribunal and is evaluated only after the accused has had a full opportunity to discredit the case against him." (Packer, 1976; 59). Thus, criminal courts outputs must be measured not just in terms of the content of decisions, but also in terms of the nature of decision making.

There is reason to believe that due process is not the norm which guides the operation of American courts. Instead of open, combative, impartial proceedings, most decisions in most criminal courts are made under closed, cooperative, partial circumstances. Herbert Packer has proposed a continuum to measure the degree to which due process guides the decisions of criminal trial courts. At one end of the continuum lies the Crime Control Model:

> The value system that underlies the Crime Control Model is based on the proposition that the repression of criminal conduct is by far the most important function to be performed by the criminal process . . . In order to achieve this high purpose, the Crime Control Model requires that primary attention be paid to the efficiency with which the criminal process operates to screen suspects, determine guilt, and secure appropriate dispositions of persons convicted of crimes . . . The model, in order to operate successfully, must produce a high rate of apprehension and conviction, and must do so in a context where the magnitudes being dealt with are very large and the resources for dealing with them are very limited. There must then be a premium on speed and finality. Speed, in turn, depends on minimizing the occasions for challenge . . . The model that will operate successfully on these presuppositions must be an administrative, almost a managerial, model. A successful conclusion to the process is one that throws off at an early stage those cases in which it appears unlikely that the person apprehended is an offender and then secures, as expeditiously as possible, the conviction of the rest, with a minimum of occasions for challenge, let alone post-audit. . . . It might be said of the Crime Control Model that, when reduced to its barest essentials and operating at its most successful pitch, it offers two possibilities: an administrative fact-finding process leading (1) to exoneration of the suspect, or (2) to the entry of a plea of guilty. [Packer, 1976: 55-59]

At the opposite end lies the Due Process Model:

> The Due Process Model rejects . . . heavy reliance on the ability of investigative and prosecutorial officers, acting in an informal setting in which their distinctive skills are given full sway, to elicit and reconstruct a tolerably accurate account of what actually took place in an alleged criminal event, . . . and substitutes for it a view of informal, nonadjudicative fact-finding that stresses the possibility of error . . . Considerations of this kind all lead to a rejection of informal fact-finding processes as definitive of factual guilt and to an insistence on formal, adjudicative, adversary fact-finding processes in which the factual case against the accused is publicly heard by an impartial tribunal and is evaluated only after the accused has had a full opportunity to discredit the case against him. . . . The combination of stigma and loss of liberty that is embodied in the end result of the criminal process is viewed as being the heaviest deprivation that government can inflict on an individual. Furthermore, the processes that culminate in these highly

> afflictive sanctions are seen as in themselves coercive, restricting, and demanding. . . . Precisely because of its potency in subjecting the individual to the coercive power of the state, the criminal process must, in this model, be subjected to controls that prevent it from operating with maximal efficiency. [Packer, 1976: 59-61]

This Crime Control–Due Process dimension may be useful for measuring the procedural element of court outputs.

The primary characteristic of crime control systems is, of course, plea bargaining, and as a consequence most research has focused upon guilty pleas as an indicator of a crime control orientation. This has resulted in the conclusion that an overwhelming majority of criminal courts in the United States are crime control systems. However, the conclusion rests on the assumption, rarely tested, that the complexities of case processing in a crime control system can be adequately measured by means of a single indicator: the percentage of defendants pleading guilty. This assumption, and the process of plea bargaining, requires more thorough examination.

PLEA BARGAINING

Plea bargaining is the means by which most criminal cases in the United States are adjudicated. In plea bargaining the defendant exchanges his/her constitutional right to a jury trial for some concession on the part of the state. Most of the participants in plea bargaining benefit from the exchange: the guilty defendant receives a less severe sanction; the defense attorney collects his fee for minimal work; the prosecutor inexpensively disposes of the case; and the judge makes no decision on which there is much likelihood of successful appeal. The only potential losers in plea bargaining are innocent defendants, and society (to the extent that sanctions reflect the defendant's bargaining skills rather than his/her need for rehabilitation/punishment/deterrence).

Despite the many nuances of plea bargaining, the practice has traditionally been measured by the percentage of cases resulting in guilty pleas. While guilty pleas may be one aspect of plea bargaining, high guilty plea rates may not be due to action on the part of the state. Instead, other factors, such as remorsefulness, desire to begin rehabilitation, the interests of the defense attorney, and so on, may influence the defendant to plead guilty when in fact no plea bargaining has occurred. Thus, while generally single indicators of any concept are undesirable, there is an especially pressing need to develop multiple indicators of plea bargaining.

Since plea bargaining may take several forms, including charge reduction, charge consolidation, concurrent sentences, and sentence leniency, several indicators of plea bargaining in the different systems must be used. Seven variables were investigated as potential indicators of system commitment to plea bargaining:

(1) the percentage of defendants pleading guilty;

(2) the percentage of defendants going to jury trials;

(3) the percentage of convictions on reduced, or lesser, charges;

(4) the percentage of defendants facing multiple charges;

(5) whether the system penalizes defendants convicted at trial;[10]

(6) a measure of the degree to which the initial charge predicts the severity of the sentence (an indicator of either "overcharging" in the initial charge or "undercharging" in the conviction charge due to plea bargaining;[11] and

(7) the severity of the sentences given.

The interrelationships of the various measures is complex, and in some instances startling (see Table 4). First, regarding guilty pleas, the data suggest that guilty pleas are not the major elements of plea bargaining. For example, high use of guilty pleas is associated with (1) *low* use of charge reductions; (2) *high* use of juries; (3) a tendency *not* to penalize innocent pleas; and (4) a *high* percentage of cases with only a single charge against the defendant.

TABLE 4
INDICATORS OF PLEA BARGAINING[a]

	1.	2.	3.	4.	5.	6.	7.
1. % Pleading Guilty	1.00						
2. % Jury Trials	.28	1.00					
3. % No Charge Reduction	.42	.06	1.00				
4. % Single Charge	.73	.02	.41	1.00			
5. Penalty for Innocent Pleas	.27	–.08	–.05	.35	1.00		
6. Impact of Initial Charge	–.07	.30	.52	–.06	–.25	1.00	
7. Sentence Severity	–.51	–.29	–.45	–.51	–.16	–.25	1.00

[a]Entries are Pearson correlation coefficients.

Further, use of guilty pleas is unrelated to the impact of the initial charge on sentencing. Thus, it seems appropriate to conclude that the frequency of guilty pleas in a system is a quite poor indicator of the system's commitment to plea bargaining.

Other patterns are also seen in the data. Use of charge reductions is inversely related to the impact of the original charge on sentences and the impact of the original charge is related to the use of juries and *not* penalizing innocent pleas. Overall, however, the pattern exhibited in the correlation is complex and not easily comprehended.

An effective means by which the complex relationships among these variables can be unraveled is factor analysis. These six indicators of plea bargaining (omitting the percentage pleading guilty) were subjected to principal components factor analysis. The results of the factor analysis are shown in Table 5.

Two dimensions of plea bargaining practices emerge from the analysis. The first factor seems to indicate whether the system manipulates sentences as part of plea bargaining. The practices of multiple charges, severe sentences, and penalizing defendants pleading innocent characterize one end of the scale, while single charges, no penalties, and lenient sentences characterize the other end. The second factor indicates the degree to which the system manipulates *charges* as part of plea bargaining. One end of the dimension is characterized by a strong impact of the original charge, the absence of charge seriousness reductions, and, to a lesser degree, high use of juries, lenient sentences, and a tendency *to penalize* innocent pleas. It is significant that the tendency *to penalize* is associated with the absence of charge manipulation because it suggests that plea bargaining may still occur, but with somewhat

TABLE 5
FACTOR ANALYSIS OF PLEA BARGAINING INDICATORS

Indicator	Factor Loadings (Varimax Rotation)[a] Factor 1	Factor 2
% No Charge Reduction	.47	.66
Impact of Initial Charge	–.06	.86
% Single Charge	.87	.02
Penalty for Innocent Pleas	.63	–.47
Sentence Severity	–.70	–.44
% Jury Trials	.02	.53
Eigenvalue	2.18	1.56
% Variance Explained	36.4	26.0

[a]Several oblique rotational alternatives to orthogonal rotation were investigated but the factor remained essentially uncorrelated (with bi-quartimin rotation r = .09). Therefore orthogonal rotation was used. Low scores on the factor scores indicate commitment to plea bargaining.

different manifestations. Interestingly, both factors suggest that plea bargaining is associated with *severe* not lenient sentences. Finally, sentence manipulation and charge manipulation are unrelated to each other (see note[a] to Table 5). Factor scores from these two dimensions of plea bargaining can be used as indicators of the propensity of the system to engage in the two forms of plea bargaining.

The relationship between the plea bargaining factor scores and the rate of guilty pleas further undermines the use of guilty pleas as an indicator of plea bargaining. Sentence manipulation is strongly related to the guilty plea rate (r = .73) but the *less* manipulative systems rely on guilty pleas *more.* Similarly, the less the system relies on charge manipulation, the greater the use of guilty pleas, although the tendency is slight (r = .18). This merely reinforces the conclusion that the guilty plea rate does not necessarily reflect state participation in plea bargaining.

Thus, the analysis suggests that Iowa criminal courts engage in two different forms of plea bargaining: charge and sentence manipulation. Neither practice is related to the guilty plea rate in the manner normally asserted. Nor is either practice related to the other, a finding which suggests mutual exclusivity.

As with the other indicators of court outputs, there is considerable variation in plea bargaining practices among the counties. These differences represent differences in the policy commitments of the different systems. The officials (judges and prosecutors) of some court systems are apparently willing to tolerate deviations from due process procedures, whereas others are less willing. While plea bargaining is certainly not the only dimension of a Crime Control orientation, it is plausible to argue that the differences among systems in their orientations toward plea bargaining are reflective of the more general Crime Control–Due Process continuum. Using the factor scores as measures of court policy, the causes of inter-system differences on this and the other policy outputs is worthy of more rigorous consideration.

CORRELATES OF POLICY OUTPUTS

Three dimensions of the outputs of criminal justice systems have been identified and measured: effectiveness, punitiveness, and commitment to due process. These are measures of the policies made by the trial courts in each of the counties. The different policy outputs are to a large degree independent of each other (see Table 6). There is some tendency for effective systems to rely less on plea bargaining, perhaps reflecting a preference for formal open procedures. Punitiveness is related to sentence manipulation, suggesting that plea bargaining is associated with more severe, not lenient, treatment of

TABLE 6
CORRELATIONS AMONG DIMENSIONS OF COURT OUTPUTS[a]

	1.	2.	3.	4.
1. Effectiveness	1.00			
2. Punitiveness	–.03	1.00		
3. Sentence Manipulation	.28	–.68	1.00	
4. Charge Manipulation	–.01	–.17	.11	1.00

[a]Entries are Pearson correlation coefficients.

defendants. This further suggests that plea bargaining may not actually benefit defendants (cf. Shin, 1973). Over all, however, three relatively independent dimensions, one with two independent subdimensions, are present in the policy outputs of these courts.

Two questions remain: why do the different systems differ in outputs, and what consequences flow from the differences? Neither of these questions can be thoroughly considered using available data, but some tentative propositions can still be advanced. Policy differences across systems, for instance, may reflect differences in the characteristics of the environments of each of the courts. Policy may reflect either the inputs of mass publics, or the constraints or exigencies arising from environmental conditions. The impact of several of these influences on policy outputs can be determined.

PUBLIC OPINION INPUTS

One possible determinant of criminal justice policies is public opinion in the counties. Counties in which public opinion is significantly more anticriminal are expected to be effective, punitive, plea bargaining oriented counties, if public opinion has any impact on the operation of the courts. From surveys conducted by the *Des Moines Register and Tribune* three measures of public opinion relevant to criminal justice policy makers have been constructed: (1) support for forced restitution by convicted criminals; (2) support for the death penalty; and (3) concern over crime.[12] While the responses to these questions may not be directly relevant to the operation of the criminal justice system, they can be taken as indicators of mass opinion on matters of crime and justice. The relationships among the public opinion variables and the measures of court outputs are shown in Table 7.

The impact of public opinion on court policies is uneven. More anticriminal counties are somewhat more effective, and somewhat more punitive, but public opinion appears to have a negligible impact on plea-bargaining practices. It is reasonable that public opinion would have the greatest impact on criminal justice policy making because punitiveness is the most salient

TABLE 7
PUBLIC OPINION DETERMINANTS OF COURT POLICIES

	Effectiveness		Punitiveness		Sentence Manipulation		Charge Manipulation	
Public Opinion	r	Beta	r	Beta	r	Beta	r	Beta
% Favoring Restitution	.30	.34	.13	–.01	.13	.20	–.00	–.03
% Favoring Death Penalty	–.01	–.11	.28	.34	–.19	–.25	.12	.11
Priority of Crime	.07	.02	.21	.28	.10	.02	–.08	–.06
R^2 (all three public opinion variables)	.10		.15		.08		.02	

dimension of court activity to mass publics. Similarly, plea bargaining, until recently a subterranean practice, is the least understood of the policy dimensions. Overall, the impact of public opinion is not great, nor is it minor. Public opinion may serve only to set broad parameters for court policy within which policy makers still have considerable discretion. These constraints differ from system to system, but because of their relative lack of specificity, they are unlikely to be able to account for much of the variance in policy outputs. This, however, does not mean that public opinion is totally unimportant.

ENVIRONMENTAL CONSTRAINTS

Policy makers may also be constrained by environmental factors in the policies they make. One factor which has particular importance is the caseload of the system. These courts differ enormously in the number of criminal cases they process, and there are theoretical reasons for expecting caseloads to be related to outputs.[13] For instance, plea bargaining is usually justified in terms of the necessity of finding efficient case-processing procedures in the face of resource shortages. The costliness of trials is said to justify state concession to elicit guilty pleas. Caseloads may also be related to effectiveness, if ineffectiveness is a result of overloading the systems with demands. Finally, because plea bargaining is related to severe sentences, high caseloads should be associated with punitiveness.

Each of these expectations is supported by the data: high caseloads are associated with lower effectiveness, greater punitiveness, and a greater commitment to plea bargaining (see Table 8). Thus, criminal justice policy does reflect the constraints imposed by environmental conditions. The data also provide a means by which previously conflicting findings on plea bargaining and caseloads can be resolved. Recent research has cast doubt on the common assumption that plea bargaining is a response to caseloads (Heumann, 1975;

Feeley, 1975; Eisenstein and Jacob, 1977: 238-239). This research has relied, however, on the guilty plea rate as the indicator of plea bargaining. Using that measure, a similar conclusion can be drawn from these data: as caseloads increase guilty pleas *decrease* (r = - .46). However, as seen in Table 8, heavy caseloads are directly associated with plea bargaining when the more sophisticated and complete measures of plea bargaining are used (as caseloads increase, plea bargaining increases). This adds additional support for the conclusion that the rate of guilty pleas is not an appropriate measure of plea bargaining. High rates of guilty pleas may occur without the state offering inducements (positive or negative) to plead guilty. A sizeable number of defendants apparently plead guilty without explicit plea bargaining, perhaps on the basis of some (unfounded) hope that the court will act less punitively (cf. Church, 1976).

The impact of several other environmental variables on court policy was also investigated and the results are shown in Table 8. The four measures of policy outputs were regressed on the relative wealth of the county,[14] the economic prosperity of the county (unemployment rate), the occupational

TABLE 8
REGRESSION OF COURT POLICIES ON ENVIRONMENTAL FACTORS

Environmental Factors	Effectiveness	Punitiveness	Sentence Manipulation	Charge Manipulation
Caseload				
r	–.36	.54	–.64	–.45
Beta	–.05	.34	–.33	–.55
Unemployment				
r	–.23	.05	–.40	.04
Beta	–.39	–.15	–.24	.21
Wealth				
r	–.46	.45	–.49	–.22
Beta	–.54	.04	–.16	.26
Political Complexion				
r	.06	.31	–.49	–.23
Beta	.18	.16	–.24	–.09
% Manufacturing				
r	–.16	.45	–.50	–.20
Beta	.18	.36	–.21	–.17
R^2–caseload only	.13	.29	.41	.20
R^2–all environmental variables	.34	.40	.61	.25

structure (percentage employed in manufacturing), and the political complexion of the counties.[15] The caseload variable was entered in the equation first because it is the most proximate cause of court policy. The results of the regression suggest several conclusions: (1) environmental influence makes a nonnegligible independent contribution to explaining variation in court policy outputs; (2) punitive court policy reflects heavy caseloads and a greater percentage of working class constituents; (3) beyond caseloads, no single environmental variable makes a significant independent contribution to explaining sentence or charge manipulation; and (4) the most effective criminal justice systems are to be found in low income, economically prosperous counties (possibly a reflection of either working class crimes being more likely to result in official court action, or working-class citizens reporting crimes at a lesser rate, thereby affecting the denominator of the effectiveness measure). One must be cautious in drawing conclusions from these data because of the small N, multicollinearity, and the highly inferential nature of the variables, but it does appear the court policy outputs are sensitive to environmental constraints. Heavy caseloads affect policy the most, but public opinion and characteristics of the jurisdiction of the court are also influential.

CONSEQUENCES OF JUDICIAL POLICY MAKING

The final question to be asked is: what impact do the policy decisions made in the criminal courts have on crime, justice, society, or politics? Rather than concluding the analysis as if policy making were the ultimate endogenous variable, it may be more fruitful to consider what impact, or consequences, judicial policy making has. In particular, we can ask if court policy has any impact on (1) the level of crime in society, or (2) public support for the court system. Both of these concerns have a direct bearing on judicial accountability.

The indicators of each of the policy dimensions were correlated with change in the crime rate (UCR estimates) in each of the counties for the years 1974 and 1975. If court policy making has any impact on the level of crime in society, then at least a slight correlation between crime and judicial policy should be observable. Realizing that there may be a lag between the adoption of court policy and any impact on society, crime rates for both 1974 and 1975 were examined. The results are shown in Table 9.

Most of the correlations between policy outputs and change in crime rates are higher for the two-year period than for the one-year period, suggesting a time lag before policies have an impact on the rate of crime. Overall, the policies do have a strong impact on change in the crime rate: over one-third

TABLE 9
THE IMPACT OF COURT POLICY ON CRIME AND SUPPORT

Policy	Change in Crime 1973-1974		Change in Crime 1973-1975		Percent "No Retain" Votes	
	r	Beta	r	Beta	r	Beta
Effectiveness	–.31	–.18	–.21	–.06	.12	.22
Punitiveness	.07	–.19	.11	–.23	.09	–.05
Sentence Manipulation	–.43	–.48	–.47	–.58	–.21	–.30
Charge Manipulation	–.11	–.13	–.35	–.37	.03	.02
R^2		.26		.39		.08

of the variance in the rates can be explained by reference to actions of the criminal justice system. The system commitment to due process has the strongest impact on crime of any of the policy variables. Systems relying most heavily on plea bargaining are the systems with the greatest increases in the rate of crime. An opposite and considerably smaller impact can be observed with regard to punitiveness: the more punitive the system, the less the increase in crime (holding all other variables constant). Finally, the effectiveness of the system is unrelated to changes in crime.

The data have produced the somewhat surprising conclusion that plea bargaining is associated with increases in crime. This is despite the fact that plea bargaining is associated with severe penalties, and punitiveness tends to reduce crime. If criminal behavior to some extent reflects the belief that no sanctions will be imposed, as recent research suggests, then plea bargaining may create the image that the system is manipulatable, and hence lead to the impression among potential criminals that sanctions would not be forthcoming. It is apparently not the severity of the sanction which has the greatest deterrent impact, but rather the certainty that the sanction will be imposed. It is not the *actual* chance of conviction which is associated with the change in crime rates (beta = –.06), it is instead the symbolic output of the system. It is the widespread perception, fostered by plea bargaining, that the system is incapable of responding effectively to alleged offenders. Due process may therefore impose efficiency costs, but, as Arnold (1935) observed 45 years ago, the impact of the symbolic outputs on compliance may well offset the loss in efficiency.

Court policy may also have an impact on the level of mass support or satisfaction with the judicial system. The attitudes of mass publics toward the judicial system may be a function of the perceived effectiveness of the courts in performing their functions (see Jacob, 1971), or mass attitudes may be due to factors totally unrelated to the performance of the criminal justice system. One way in which this question can be evaluated is to examine the relationship between court policy and the level of dissatisfaction with the courts as reflected in voting returns for criminal court judges. Voting statistics may be a poor indicator of dissatisfaction because they may reflect a multitude of idiosyncratic influences. On the other hand, judicial elections in Iowa are retention elections, elections which tend not to reflect policy and other concerns. They can legitimately be interpreted as plebiscites on the court system as a unit rather than as attitudes toward particular judges. Using the percentage of the vote opposing retention of the judge or judges, it is possible to determine if court policy affects mass satisfaction with the courts.

There is only a slight relationship between the two variables (see Table 9). There is some small tendency for systems not relying on plea bargaining to draw more electoral support, but more effective systems tend to receive *less* support. Thus, court policy may well have an influence on the amount of crime in society, but it has little impact on mass attitudes toward the courts. Mass attitudes probably instead reflect generalizations from personal or vicarious experiences which bear little relationship to the actual performance of the criminal justice system.

CONCLUSIONS

This research was designed to measure the policy outputs of criminal courts in Iowa. Three dimensions of the outputs were identified and indicators of policy outputs were constructed. It was then determined that court policy to some extent reflects external environmental pressures, including public opinion, but that the exigencies imposed by caseloads appear to be the single most important determinant of policy outputs. Finally, court policy has a significant impact on the amount of crime in society, but no impact on the level of support the court system receives.

The analysis, of course, can offer no direct conclusions about the question of judicial accountability. Indeed, because many of the measures employed in the research are crude at best, the findings of the analysis are especially tentative. Nevertheless, the analysis is suggestive of some important questions for further accountability research:

(1) Accountability to whom: the analysis suggests that mass publics are not accurately evaluating judicial policies. This is not surprising in that mass publics probably do not accurately evaluate the policy outputs of any institution. If not to mass publics, to whom should judicial policy making be accountable? Certainly, accountability to private organizations such as the ABA is not the solution, because these organizations are even more unaccountable than the system itself. The only other groups to which the courts might be held accountable would be other elite groups, perhaps even other

segments of the political system. This in turn suggests a model of plural elites making conflicting demands on the judiciary, which raises the question of politicization of the courts.

(2) The politicization of the courts: perhaps one way in which courts might be held more accountable is by shedding the cloud of objectivity and impartiality which enshrouds the judicial process. The most effective way in which policy accountability can be avoided is to deny that policy is being made. Yet this research has demonstrated that there is a strong influence of court policies on the level of safety in society, a vital issue of public policy. If policy is ever to be controlled, it must first be acknowledged. This would require that the judiciary become politicized—that is, that the courts openly assume a central position in the political arena. For instance, rather than special procedures for the selection of judges, procedures which have the effect of minimizing mass participation (e.g., nonpartisan, spring elections), judges could be selected in exactly the same way in which other policy makers are selected. Judges should be allowed and encouraged to campaign on the basis of policy positions. How a judge will (or has) sentence those who break marijuana laws, for instance, is equally as essential to accountability as how a legislator will (or has) vote on decriminalization. Regardless of how effective procedures for accountability are, the whole issue can only be addressed if judicial decisions are treated fundamentally as authoritative allocations of values for the political system. The choices are essentially to deny that policy making occurs and thus provide no means of control, or to accept the political influences on judging and try to control them.

(3) Accountability for what: this research suggests that *how* courts make policy is at least as important as *what* policy is made. Accountability thus concerns more than simply the substantive policy outputs of courts: symbolic outputs are also quite important. This, however, raises a conflict: presumably, courts are expected to have some impact on the amount of crime in a city, but expectations concerning how the courts should operate (i.e., due process expectations) may impede the effectiveness of the system. This creates a need for the system to operate with duplicity: bail amounts, for instance, are set at very high figures largely to preempt potential critics, while the actual costs to the defendant of bail is one-tenth of the face amount. Accountability can thus create severe strains on organizations, and may impede effectiveness.

(4) Accountability and minorities: proponents of accountability operate from a majoritarian bias. Given the elitest nature of the judiciary, and its special role as a protector of minority rights, accountability may mean majority tyranny. To have required, for example, that the U.S. Supreme Court be accountable in the last 25 years would have drastically curtailed the civil rights movement.

The questions of the possibility and desirability of judicial accountability must for the present remain open. We should therefore be cautious about making facile assumptions and conclusions without more empirical evidence concerning the levels, causes, and consequences of accountability.

NOTES

1. There is no better example of this than the experiences of the Warren Court in the 1960s: it was not the sudden increase in policy making under Warren which brought about mass dissatisfaction with the Court, but rather that the policies made were out of step with majority opinion on so many issues at once. The Court was also relatively candid about the policy nature of its decisions. By the same token, the change in public perceptions of the Court under Burger certainly has not resulted from a decline in policy-making activity.

2. This can also be seen in the type of "reforms" in judicial selection which have been implemented in the last decade. The trend is unmistakably toward less "political," and hence less accountable, methods of selecting judges (e.g., The Missouri Plan).

3. Discretion is also central to the question of accountability because accountability basically concerns the proper or improper exercise of discretion.

4. This is the central question for many judicial scholars. See, for example, Jacob (1978).

5. Consider, for example, the ease with which southern federal district court judges were able to defeat appellate court policy on school desegregation in the 1950s and 1960s. See Peltason (1971).

6. Some additional error may be introduced by the fact that the date on which the crime was committed is unknown. Instead, the arrest date is used as the estimate of the year in which the crime was committed.

In addition, the three counties in which court records were sampled are excluded from this portion of the analysis to insure that sampling error does not influence the results.

7. The clearance rate is an indicator that "law enforcement agencies . . . have identified the offender, have sufficient evidence to charge him and actually take him into custody. Crime solutions are also recorded in exceptional instances when some element beyond police control precludes the placing of formal charges against the offender, such as the victim's refusal to prosecute after the offender is identified or local prosecution is declined because the subject is being prosecuted elsewhere for a crime committed in another jurisdiction" (FBI, 1974:42).

8. The impact of the seriousness of the charge on which the defendant was convicted on the severity of the sentence imposed has been partialled out by means of standardized Z scores. For further details on the process see Gibson (1978).

9. The impact of charge seriousness on the amount of bail has been removed. The amount of bail was regressed on the seriousness of the charge resulting in the following equations: $\hat{Y} = 11631.8 - 656.2 *$ charge seriousness. The measure of bail used is simply the actual amount of bail–Y. It was not necessary to remove the impact of charge seriousness on pretrial incarceration because the correlation with charge seriousness is so slight.

10. The measure of whether the state penalizes innocent pleas is a dichotomy constructed on the basis of a t-test between the average sentence severity of convictions resulting from pleas compared to the average sentence severity of convictions resulting from trial. Because of the small number of trials in some counties, only 14 of the 25 counties could be assigned scores on this measure. For analysis purposes, pairwise missing data deletion is used.

11. For many of the court systems there is a very significant nonlinear impact of the original charge on sentences, as reflected by the fact that eta^2 is considerably larger than r^2. The linear impact is considered to be the most appropriate measure, however, because the legal effect of charge seriousness should be linear (or at least monotonic). To control for the nonlinear effect would be to eliminate variance more properly attributed to the policy making of the various court systems.

12. For further information on the surveys, the questions, and the techniques used to score each of the counties, see Gibson (1975).

13. There is a serious multicollinearity problem with the measure of caseloads: caseloads, urbanization, crime, and court budgets are all related very strongly. Consequently, their relative impact on court policy cannot be evaluated. The discussion and analysis assumes that the primary impact of this syndrome of variables is through caseloads, since caseloads are the most proximate and the most theoretically relevant of the variables.

14. Wealth is measured by a factor score representing several census variables. See Gibson (1975).

15. This is a factor score resulting from a factor analysis of vote returns for 1972 statewide elections. For further details, see Gibson (1975).

REFERENCES

ARNOLD, T. (1935). The symbols of government. New Haven, Conn.: Yale University Press.

BEISER, E. N. (1972). "Lawyers judge the Warren Court." Law and Society Review, 7:139-149.

CASEY, G. (1974). "The supreme court and myth: an empirical investigation." Law and Society Review, 8:385-420.

CHURCH, T. W., Jr. (1976). "Plea bargains, concessions and the courts: analysis of a quasi-experiment." Law and Society Review, 10:377-402.

EISENSTEIN, J. and JACOB, H. (1977). Felony justice: an organizational analysis of criminal courts. Boston: Little, Brown.

Federal Bureau of Investigation [FBI] (1974). Uniform crime reports for the United States. Washington, D.C.: U.S. Government Printing Office.

FEELEY, M. (1975). "The effects of heavy caseloads." Paper presented at the 1975 Annual Meeting of the American Political Science Association, San Francisco.

FRANKEL, M. E. (1973). Criminal sentences: law without order. New York: Hill & Wang.

GIBSON, J. L. (1978). "Judges' role orientations, attitudes and decisions: an interactive model." American Political Science Review, 72.

--- (1975). "The judicial decisions: social and psychological determinants of the decisions of criminal courts in Iowa." Ph.D. Dissertation, University of Iowa.

HEUMANN, M. (1975). "A note on plea bargaining and case pressure." Law and Society Review, 9:518-524.

HOROWITZ, D. L. (1977). The courts and social policy. Washington, D.C.: The Brookings Institution.

JACOB, H. (1978). Justice in America: Courts, lawyers, and the judicial process. (3rd edition) Boston: Little, Brown.

--- (1971). "Black and white perceptions of justice in the city." Law and Society Review, 6:69-89.

PACKER, H. L. (1976). "Two models of the criminal process," in George F. Cole (ed.), Criminal justice: law and politics. North Scituate, Mass.: Duxbury Press.

PELTASON, J. W. (1971). Fifty-eight lonely men (2nd ed). Champaign: University of Illinois Press.

POPE, C. F. (1975). "Offender-based transaction statistics: new directions in data collection and reporting." Washington, D.C.: U.S. Department of Justice.

SHIN, H. J. (1973). "Do lesser pleas pay: accommodations in the sentencing and parole process." Journal of Criminal Justice, 1:27-41.

SKOGAN, W. G. (1974). "The validity of official crime statistics: an empirical investigation." Social Science Quarterly, 55:24-38.

WILDHORN, S., et al. (1976). Indicators of Justice: measuring the performance of prosecution, defense, and court agencies involved in felony proceedings: analysis and demonstration. Santa Monica, Cal.: The Rand Corporation.

6

Accountability of the Courts

STEPHEN L. WASBY

□ IN THE 1970s, THE FREQUENCY of attacks on the courts, particularly the federal courts, for exceeding their authority, have increased. With public opinion toward civil rights turning conservative, the cry of "government by the Judiciary" heard in the 1930s and earlier when courts overturned legislation regulating the economy and, more recently, in the 1960s during the Warren Court's "civil liberties revolution" is again being voiced (Berger, 1977; Graglia, 1976). The moderate-to-conservative position of the Supreme Court under Chief Justice Burger, particularly in limiting access to the courts by those wishing to challenge government actions, has provided a model against which some court-watchers evaluate the "activism" of other judges. Judges have gone too far in dealing with problems such as deficiencies in state mental health and corrections systems which should properly be the domain of other branches of government; furthermore, we are told that the courts lack the capacity to deal with such problems (Horowitz, 1977; but see Wasby, 1978b).

Such criticisms, usually based on the assumption that the courts, not those whose actions or inactions the courts are attempting to correct, are acting improperly, are often only thinly veiled polemics against the courts' liberal results. Thus Judge Garrity's rulings on school desegregation in Boston (*Morgan* v. *Hennigan,* 1974 et seq.), well within established constitutional law, are criticized, while the Boston School Board's adamant refusal to obey clear state law is played down or ignored. Similarly, it is Judge Johnson's orders (*Wyatt* v. *Stickney,* 1971 et seq.), not Alabama's unwillingness to

AUTHOR'S NOTE: *I wish to express my appreciation to John H. Baker and Ronald Mason, Southern Illinois University at Carbondale, and James L. Gibson, University of Wisconsin–Milwaukee, for their comments on an outline and earlier drafts of this chapter.*

correct long-standing appalling mental hospital conditions, which are attacked. That the judges often arrive at their conclusions through legitimate interpretation of the spate of civil rights and social policy statutes passed by Congress starting in the 1960s is also often conveniently ignored.

One effect of attention paid to seemingly far-reaching judicial rulings is continuing deemphasis of the largely routine and unproblematic actions of most trial courts, particularly at the state level, despite the immense policy effects of those courts' rulings in the aggregate. Greater attention has been drawn to such courts by the growing literature on the behavior of the criminal courts (e.g., Eisenstein and Jacob, 1977). However, they continue to operate at the low level of visibility produced by the disproportionate emphasis that the "upper-court myth" has led us to place on appellate courts in general and the U.S. Supreme Court in particular.

Heavy emphasis on ground-breaking decisions and on the higher courts produces a view, reinforced by the "imperial judiciary" arguments, that courts are institutions which are not and cannot be accountable, to the broad political community, that they cannot be restrained from exercising unfettered discretion. Further reinforcement of such a perspective comes from the myth of judicial independence and from statements by judges themselves, of which Chief Justice Hughes' remark that "the law is what the judges say it is" or Justice Stone's comment that "the only check on our own exercise of power is our own sense of self-restraint" (*U.S.* v. *Butler,* 1936:94) are representative.

We also usually look at courts in terms of how they hold other institutions (legislatures, executives, regulatory agencies) accountable for their actions. Judicial review of actions taken by other elements of the government is essential if we are to have a "government of laws not of men only." Yet to focus solely or predominantly on such judicial action makes it appear that the courts exist in their own world to control others, without themselves being controlled. Adopting a different view is, however, possible; emphasis can be placed on the accountability *of* the courts. Attention to local trial courts has assisted in adoption of such a perspective but it does not go far enough. The tendency has been to criticize such courts for failing to act independently of local political interests instead of to stress ways in which connections between local courts and such interests help make the courts accountable.

There is no reason why we cannot adopt a perspective which stresses accountability, at least to see where it might lead us. That means of accountability appear to be few and judicial independence substantial does not mean that accountability does not exist—only that we need to search harder to find it. That the means to achieve accountability which we might identify are often less than effective so that courts are not fully accountable also does not mean that no accountability takes place; indeed no governmental institution is fully accountable. Keeping in mind such a comparison of the courts' accountability with that of other institutions might produce a different and more favorable picture than if one were to focus solely on the courts.

The following is a preliminary exploration of various ways in which courts are, or might be made, accountable. Because the topic of judicial accountability has not previously been given much attention, little has been written about the courts with accountability as the focus, but many studies of courts and the judicial process can appropriately be brought to bear on the subject to show both how accountability might be achieved and how the mechanisms might not operate effectively to achieve accountability. So as not to exclude possible means of accountability or relevant materials, "accountability" is used loosely. Generally, we use accountability here to mean keeping an institution's decisions in line with community political or social values and otherwise imposing constraints on the courts' exercise of discretion. The latter half of the definition makes it appear a form of influence. In a sense, accountability and influence are complementary aspects of viewing the same process; any person or institution exercising influence over the courts *is* contributing to the courts' accountability, at least to that person or institution.

From the point of view of democracy, accountability would be most complete if the political or social values toward which an institution's decisions are drawn were those of the entire community. However, it is also possible that an institution will be found accountable to only a portion of the political community, whether its elites or its working class. In this sense, *whether* there is accountability is a different question from the one of *to whom* accountability exists.

Identifying mechanisms of accountability is one important task undertaken here. To help in arranging available materials, the mechanisms may be categorized. We use two broad tentative categories here, the legal system and the more broadly defined political system. They roughly parallel the legal subculture and the democratic subculture identified by Richardson and Vines, who feel the influence of the former is clear while "the influence of the democratic subculture is somewhat less visible" (Richardson and Vines, 1970:10). Accountability within the legal system includes judges' socialization to lawyer and judge roles, the constraints which precedent and reversal impose on judges, and the judiciary's organizational needs. Political accountability includes the effect of the political environment, including public opinion, on judges, and noncompliance with and resistance to disliked decisions. Mechanisms of judicial selection and removal fit into both categories used here, although Richardson and Vines place rules and norms on the recruitment of judges in the legal rather than democratic subculture.

No potential mechanism of accountability operates in isolation. As we will see, some mechanisms serve to reinforce each other, while others may seem to work at cross-purposes. While we return to this point in concluding, it should be noted that there is at least a potential conflict between accountability within the legal system and political accountability, that is, accountability to the entire political community or some of its elements. To the extent the courts remain primarily accountable to lawyers, particularly the organized bar, there is the possibility that competing values held by the larger, non-lawyer community will be correspondingly discounted. If this be so, the predominance of lawyers in the accountability of courts may be viewed as a special case of professional organizations' avoidance of political accountability to the broader political community. Other professions, e.g., physicians, have generally been able to reduce accountability by keeping medicine a largely private social institution, although they are becoming increasingly subject to public scrutiny as public financing of health care (Medicare, Medicaid) increases. Lawyers, by contrast, have sought to retain control of an institution–the courts–which is formally public but which lawyers (and judges) have kept largely private, as we shall note later in discussing the courts' relative lack of openness.

ACCOUNTABILITY WITHIN THE LEGAL SYSTEM I: SOCIALIZATION

One ingredient of judges' accountability is the socialization which, like any other role occupants, they have undergone. In one sense, judges are strongly socialized; in another, their socialization is weak. Their extensive socialization is as lawyers and comes through their law school training and the years they spend practicing law, with their basic outlooks heavily conditioned by their early training. In this manner, they learn, among other things, to pay attention to precedent and to obey higher courts. Extensive trial practice may serve to socialize them to other norms such as cooperation with other regular courtroom practitioners and the greater immediate importance of the trial judge, but earlier training is hard to displace and tends to follow lawyers when they assume the judicial robes. Indeed, those with the usual limited means of exposure to the judiciary and without intimate knowledge of the judicial system derived from direct experience of close (family) ties to judges may, when they become judges, operate on the basis of a conventional view that precedent should be closely followed. One might infer this to be the case from the finding that Supreme Court justices without prior judicial experience were less likely to depart from precedent than were those with such experience; similarly, justices not from families containing judges were more likely to follow precedent than were those from "Judicial families" (Schmidhauser and Gold, 1962). Those exposed to the judiciary more directly act with more flexibility because they have learned the realities.

In addition to their socialization to norms of how to decide cases, lawyers are also socialized to the norm of judicial independence, the idea that judges ought not be accountable to the "political" branches of government. To urge that judges' decisions be based on precedent is to turn judges away from the legislative and executive branches and inward toward their colleagues. More, however, is involved: lawyers are taught that even when applicable precedent is unavailable, the sources of judges' decisions should be found in "the law" and not in popular feelings about the topics on which they are called to rule. Thus lawyers are socialized to norms about accountability itself, reinforcing the myth of judicial independence. Moreover, when matters of legal ethics are discussed, as in courses on Professional Responsibility, lawyers learn that the source of ethical rules—for lawyers and judges alike—is not the broader community but the organized bar, to which lawyers should consider themselves accountable while judges should pay particular attention to other judges for determinations of what is appropriate conduct.

If most judges' socialization as lawyers is strong and does contain inculcation of norms (concerning precedent and judicial independence) relevant to judicial behavior, their explicit socialization to specific judicial roles is far less extensive. Little, other than studying cases and perhaps some courtroom observation, occurs before a person becomes a judge. American judges are lawyers who *happen* to become judges without any particular focused preparation or specific training as judges. This is very different from the pattern in certain European countries, where the decision to receive training as either lawyer *or* judge is made quite early, with a long, systematic apprenticeship in various lawyer and judicial positions required for the former. One can view American lawyers' law practice as an apprenticeship for judging, but it is seldom purposefully so and is at best indirect, with no direct experience required in judicial or quasi-judicial positions. (The American Bar Association strongly prefers judges to have had trial experience as lawyers, but that is far short of more direct experience.) Most important, American lawyers' "apprenticeship" to be judges is not undertaken as part of a definite career planned from the beginning. Nor is there any particular time at which one becomes a judge: some do it early, while others do it later almost as a way of retiring from a more active life in legal practice.

We have begun to train judges *as* judges only quite recently. Such training still is not part of the selection process; it does not occur until after judges are selected, when the seminars of the Federal Judicial Center or the National College of the Judiciary become available. What is provided, while considered invaluable, is also limited in duration. The result is that the most important elements of judges' direct socialization to their judicial positions comes from contact with the judges with whom they serve; reliance on each other for such learning is reinforced by norms against consultation with nonlawyers. Even if one is a member of a collegial court, judging is essentially a solo task. This serves to limit the extent of contact with one's colleagues. Yet informal interaction, for example, at lunch, can be an effective way of communicating norms both for trial judges and appellate judges (Wasby, 1978a). Thus federal judges seeking answers to questions bothering them about how to perform their tasks, engage in considerable communication by letter and telephone as well as taking advantage of every opportunity for face-to-face communication (Carp, 1972; Carp and Wheeler, 1973; Cook, 1971). Sometimes they organize more purposeful sessions. One example is the sentencing councils held by some federal district judges, which bring them together to discuss proposed sentences in order to reduce their range for comparable offenses and offenders; less frequently held sentencing institutes serve the same function (Frankel, 1973). While such mechanisms do not produce uniformity, they do help develop some accountability to workgroup norms.

We cannot be certain of the specific effect of such socialization. However, it seems clear that existing norms of how the judicial task ought to be done will be strongly transmitted, making it particularly difficult to get judges to change their ways of operating. It also seems clear that accountability within the judicial profession rather than to others will be encouraged if not formally enforced through such socialization. If norms other than the current ones or new perspectives, for example, toward managing courts, are to be implemented, more training must be done as soon as judges enter judicial positions, when they are particularly seeking knowledge as to how to do their jobs. Habits from law school and long years of lawyering are unlikely to be overcome completely by such training, but it can have definite—and positive—effects.

ACCOUNTABILITY WITHIN THE LEGAL SYSTEM II: PRECEDENT AND OPENNESS

America's legal system is based in large measure on common law. In such a system, judges are expected to look to precedent as the basis for their rulings; *stare decisis* is supposed to constrain their actions, placing on their shoulders both the hand of the past and the weight of contemporary appellate courts. Do precedents hold judges in check? Some would respond cynically that judges use precedents only for rhetorical purposes, to rationalize decisions based on personal values, what Holmes called their "inarticulate major premises." Yet even if one rejects such a position, precedent's impact may be limited, making it less than fully effective as a mechanism of accountability. Indeed, as a "fact-skeptic" would argue, the finder of fact in a case, be it judge or jury, has much leeway in determining elusive facts within the confines of clear and controlling precedent. Even if facts are clear, judges can make many fact-choices, deciding which facts are relevant. If, as a "rule-skeptic," one puts facts to one side, there are many cases in which precedent is far from clear. If lawyers discourage their clients from bringing cases in which the law is clear, those cases which are brought to court will be disproportionately those where no (clear) precedents exist or ones in which conflicting precedents leave judges with the determinative choices.

The absence of applicable precedents in developing legal situations means a reduction in appellate courts' ability to control the lower courts. Because of the limited numbers of cases it decides, the United States Supreme Court cannot establish controlling doctrine for all issues the lower courts must face. Even the lower appellate courts, which hear a far higher volume of cases, may be hard put to issue controlling precedents for the trial courts, particularly in rapidly changing areas of the law. For example, state supreme courts did not deal with a large proportion of relevant search law questions after the Supreme Court applied the exclusionary rule for improperly seized evidence to the states; as a result, many lower courts did not have answers to questions with which they regularly would have to deal (Canon, 1973).

Precedent is also expected to play a part in the process by which judges arrive at the reasoned decisions we expect of them. "We contain our judges by method, and demand justification of their results by reason." When we say we want our judges to act neutrally, we mean "not so much the content of a decision, and certainly not its impact, but the style of operation" (Dixon, 1976:73, n.134). In particular, we wish judges' rulings to have attributes subsumed under the heading of *formality;* this is what gives legal judgments their special character:

> If . . . a *legal* solution is sought to the problems that arise from breaches of etiquette or other disruptions of the patterns or norms of social behavior, then *time* must be taken for *deliberate* action, for *articulate definition* of the issues, for a decision which is subject to *public* scrutiny, and which is *objective* in the sense that it reflects an explicit community judgment and not merely an explicitly personal judgment. [Berman and Greiner, 1972:26]

Furthermore, we insist on open trials, a norm reinforced by the Supreme Court's rulings against "gag orders" and its other rulings on the media's right to report court proceedings (*Nebraska Press Association* v. *Stuart,* 1976; *Cox Broadcasting* v. *Cohn,* 1975).

Shelves of volumes of published appellate opinions may make us forget that the judicial decision-making process is in many important ways not public. In the first place, many judicial opinions are not written and many of those which are, are not published. Few trial courts issue written opinions; even on the federal level, relatively few district court decisions are published. Proportionately more appellate rulings are published; however, an increasing number of U.S. Court of Appeals decisions—over 50% in at least one circuit—are being designated "Not for Publication." Such rulings, which often contain only a brief statement of the facts of a case and an even briefer explanation of how the court reached its conclusion (sometimes only "Affirmed," with one or two citations) are available to the parties, but they are not broadly available nor may they be cited as precedents in later cases.

More important is the fact that the process by which the courts reach their decisions is not public. One can watch a trial or appellate oral argument, but only in person because of the exclusion of the camera from the courtroom in the federal courts and in all but a few states. However, in no event is the public privy to the trial judge's deliberations, which occur in the privacy of his own thoughts, or to appellate courts' judicial conferences, which are closed even to other court personnel. Moreover, many judges take a dim view of openness, some even "running for cover" when regular court-watchers from public interest groups appear. (Would you believe moving court into the judge's small chambers as an avoidance device?) Some courts have not been quick to explain their operating procedures, and the provisions of the Freedom of Information Act do not reach federal court internal documents; in releasing information, the Administrative Office of the United States Courts goes to some lengths to mask judges' identities so they cannot be matched with particular decisions. One certainly does not have the accountability which public appearance or press conference forces on other public officials; instead judicial rulings are supposed to speak for themselves—and to say virtually all there is to say.

These comments lead to the conclusion that one should not rely heavily on precedent or openness to produce judicial accountability through reducing judicial discretion. At the same time, however, one should not too hastily underestimate the effect of expectations that judges should act in certain ways. Those who believe in role theory will understand that judges are likely to believe that they *should* follow precedent and that their opinions *ought* to be reasoned and should present the true bases of their decisions. Such beliefs, while not producing complete accountability, will not be without effects.

ACCOUNTABILITY WITHIN THE LEGAL SYSTEM III: REVERSAL

Higher courts can limit lower courts' freedom of action by establishing precedents. Accountability, again the sense of limiting discretion, may also occur through higher court reversal of lower court decisions; reversal is "the major sanction available to higher courts in their relations with subordinate judges" (Baum, 1978). We are told that judges do not like to be reversed; they will avoid it because it is embarrassing. And socialization to the norm of following rules laid down by higher courts is often quite purposeful. Federal district judges are asked to sit with the courts of appeals, not merely to assist with caseload, but primarily so they can learn how appellate courts work and what they expect of trial judges. If this socialization is effective, the appellate court's burden will obviously be decreased. Appellate court review of lower court decisions is in fact limited in several ways. Those courts with discretionary jurisdiction, like the U.S. Supreme Court with its power to grant or deny *certiorari,* cannot possibly take all cases which appellants would like them to hear. More important, the opportunity for review is eliminated in many instances because not all trial court decisions are appealed. Indeed, because the decision to appeal is in the hands of the lawyers for the parties, they play a large part in holding judges accountable; reversals cannot occur without cases they appeal.

Other limits exist on accountability of lower courts to higher courts. When appellate courts do review cases, certain rules limit their ability to overturn lower court rulings. Among these are the "clearly erroneous" test for factual matters, the "plain error" rule, and the "abuse of discretion" doctrine. Furthermore, because higher courts depend on lower courts to carry out doctrines they have developed, they cannot afford to antagonize the lower courts through frequent reversal. The result is that appellate judges normally defer to their lower court brethren: the actual rate of reversal is not high. Thus the *fear* of reversal—its symbolic effect—has to be more effective than reversal itself.

Not all lower court judges fear reversal, however: there are resistant judges who accept reversal as a necessary price for their actions, even if it lowers their reputation in the legal community. Not only is there an extreme case like that of Judge Willis Ritter of the District of Utah, whose behavior is apparently the result of erratic temperament seasoned with a belief that the Mormon Church is conspiring against him. There are also other judges frequently reversed not because of their idiosyncratic behavior, but because they do not "get the message" and insist on adhering to a certain ideology. Opponents of the Supreme Court nomination of Judge G. Harrold Carswell pointed out that Carswell had been reversed by the U.S. Court of Appeals for the Fifth Circuit more frequently than all but a half-dozen of the circuit's more than 60 district judges; that two of the others were from Mississippi indicates the part racial ideology plays in judicial behavior (and the accountability of the Mississippi judges to the local environment).

In addition, there are many lower court judges who do not directly invite reversal but who apply appellate court rulings in an extremely narrow fashion and who with some frequency openly criticize the higher courts (Canon, 1974). The demeanor of many criminal trial judges also effectively communicates their dislike of having lawyers raise or pursue motions to suppress evidence because of improper searches, thus undercutting their judicial superiors' rulings on the law of search and seizure. Such judges may receive considerable attention, but we must remember that most lower court judges do what is expected of them. This occurs because they have been socialized to do so, and whether or not they dislike reversal, most higher court rulings fall within their "zone of indifference" (Baum, 1978). From time to time, lower court judges' preferences and interests may run in a direction opposite to higher court policy, particularly in new areas of the law, but in the larger picture this occurs relatively seldom.

Among other reasons why higher courts cannot fully control lower courts is that while judicial systems may on paper look like hierarchies, they are *not* bureaucratic systems in which decisions are communicated quickly and clearly through channels from higher court to lower courts (Murphy, 1959; Baum, 1976). Appellate rulings are not orders to lower courts except for the court from which the appealed case came. As in any bureaucracy, information passing through channels becomes distorted and indeed may not come immediately to the attention of many lower courts. In fact, precedents are usually brought to judges' attention by lawyers arguing later cases; the judges, who often have little time to read rulings from other jurisdictions (see Beiser, 1972), are largely dependent on lawyers doing this and doing it accurately. Judges can, of course, do research on their own, particularly if they have law clerks, but outside the federal court system, that is not a resource provided to trial judges.

Lest lower court latitude vis-à-vis the higher courts be overemphasized, we ought to note efforts made to decrease that latitude. One reform presently under serious consideration for the federal courts is appellate review of sentencing. ("Flat sentencing" plans now being adopted in some states of course limit judges' discretion but are not aimed at lower court relations with higher courts.) The current proposals stem from the fact that trial judges' sentencing, in which they have considerable discretion, is effectively beyond appellate review; moreover, judicial review of sentencing is at best indirect, when a convicted defendant appeals a conviction because of displeasure with the sentence. What such proposed reforms perhaps indicate is that if judges in the aggregate are thought not to exercise sufficient self-restraint on matters like sentencing, systemic changes will be sought to limit them and to achieve greater accountability.

ACCOUNTABILITY WITHIN THE LEGAL SYSTEM IV: ORGANIZATIONAL NEEDS

Judicial accountability from above has obvious limits. It is, however supplemented by lateral or horizontal accountability to work-group norms. This takes place principally through the already noted socialization of judges by their colleagues. For appellate judges, horizontal accountability is primarily to other judges. For trial judges, however, others are included: the prosecutors, defense attorneys, police, and court personnel such as clerks and bailiffs who constitute the "courtroom workgroup" (Eisenstein and Jacob, 1977). Studies of criminal courts, particularly those courts with a large volume of cases, have shown that organizational dynamics, rather than justice or the need to comply with precedent, control many of the dispositions reached in those courts.

The need of the "courthouse regular" private defense attorney or the public defender to maintain ongoing relationships with the prosecutor, and the need of both prosecutors and defense counsel to appear before the same judges on a recurring basis, cannot help but affect the results in cases. Indeed, it may be other actors, not the judge, who dominate the proceedings, with the prosecutor playing a particularly active role (see, for example, Cole, 1970). Judges depend on them for information, for recommendations on bail, the acceptance of guilty pleas to reduced charges, and sentencing, and, in general, for the prompt processing of cases. In a way, this makes judges accountable to prosecutors. Yet because judges must place the final seal of approval on the dispositions prosecutors desire, one can more appropriately speak of their mutual accountability; because of mutual dependence, each member of the courtroom workgroup is accountable in some measure to each of the others.

The jury is another part of the courtroom environment. Juries may be a limited constraint on judges' actions because criminal jury trials are infrequent and because juries play a decreasing role in civil cases as well. Yet there is a right to a jury trial and, as an alternative fact-finding mechanism, juries can prevent judges from totally dominating cases. Judges rule on evidentiary points and charge the jury on the law; they can refuse to let a case go to the jury, can direct a verdict or a judgment n.o.v., and lower damage awards. Much, however, remains in the jury's hands. Where a general verdict is returned, the jury need not explain or justify its decisions, making its rulings for any reason or no reason. When their view of what is proper leads to a different result from what the law seems to require, they can decide accordingly. However, they are "at war" with the law, not across the board, but primarily in certain areas of the law (e.g., sumptuary legislation), and seldom act in maverick fashion when the weight of the evidence and the direction of the law are clear. Indeed, judge and jury are usually in agreement as to a case's result (Kalven and Zeisel, 1966). Thus they do not often force the judge to exercise his authority by returning unreasonable or outrageous verdicts. Moreover, contrary to the conventional wisdom that they are incapable of understanding complex legal formulations, juries do appear to pay close attention to the judge's instructions on the law: when those instructions are changed, results shift (Simon, 1967). If these findings suggest that juries do not often constrain judges, nevertheless they must be seen as another part of the "mix" of elements potentially producing some judicial accountability.

If organizational needs produce accountability of judges to lawyers, and, in times of crisis, political accountability, that is, movement toward community values, one should also note that, through court reform, judges have acted to decrease their dependence on lawyers by increasing their authority and capacity to control trials (Baar and Baar, 1977:100-101). Procedures for judicial control of pretrial activity in cases reduce dependence on lawyers, thus increasing judges' discretion (a person with better control over the flow of information increases his or her freedom of action). Establishing unified court systems allows judges more flexibility in controlling that system, while court administrators provide the managerial support necessary for such control; furthermore, a separate Administrative Office of the Courts (AOC) decreases dependence on the executive branch and increases leverage vis-à-vis the legislature. Ironically, to establish unified court systems and administrative offices, judges have to go to the legislature, meaning short-run political accountability to obtain longer-run decreased dependence; however, continuing battles in some states to eliminate AOCs indicate that political accountability does not disappear, although it does decrease.

Activity in the criminal courtroom is affected by more than the need to get along in order to process caseload. Interaction over time can produce a convergence in values. However, some congruence may exist from the beginning because most courtroom workgroup members are likely to come from the same legal and political culture, from the same (local) law schools, the same political associations, and even the same neighborhoods. The values shared, which often operate to limit the effect of higher court rulings, become particularly dominant in crisis situations. During major civil disturbances, some judges seem to suspend totally whatever independent judgment they have, acting like extensions of law enforcement agencies at least until calm returns (*Criminal Justice in Extremis,* 1969; Balbus, 1973). Whatever this says about judges' failure to maintain their independence when it may be most necessary, it clearly indicates an implicit accountability to the values of the immediate area.

LEGAL AND POLITICAL ACCOUNTABILITY MIXED: SELECTION AND REMOVAL

Judicial selection is one of the most important ways in which judicial accountability might be achieved. Although judges once chosen might—and do—go their own way, the more effective the screening and selection process in identifying judges with desired characteristics, the less we might need to use systems to achieve continuing accountability. A related consideration is that the method of selection helps determine the groups to which judges will attempt to hold themselves accountable, the reference groups from which they will draw cues. The basic question is whether they will be more responsive to legal-professional interests or to other constituencies. If most lawyers and judges share an orientation toward certain behavioral norms, the "merit system" method of selection will further magnify them, while they are more likely to be attenuated by partisan election systems.

Lawyers have strong preferences as to how judges should be selected. At the most basic level, there is an insistence that judges be lawyers. This sets up a primary orientation away from the general public and toward the community of "bench and bar." Whatever one may think of lay Justices of the Peace (JPs), they are more like "the rest of us" than lawyers; because they are more fully embedded in the local community, they are more immediately accountable than lawyer-judges. Having achieved a virtual monopoly of judicial positions in most states, lawyers prefer that control of selection and removal remain within the legal community; if it cannot, the organized bar prefers that partisan influences remain at a minimum. Stressing the norm of judicial independence, lawyers frown on situations requiring judicial candidates to campaign; indeed, these preferences have been embodied in Codes of Judicial Conduct.

The bar is aided in its quest to play as large a role as possible in judicial selection, regardless of the selection method used, by the public's willingness to accept professional opinion about matters related to the profession. Increasing public skepticism with respect to the closely interrelated matters of advertising by lawyers, lawyers' fees, and the delivery of legal services to the middle class does not seem to carry over into judicial selection. Considerable importance is attributed to bar association ratings. This is most evident in the acceptance of the "Qualified"/"Not Qualified" evaluations of the American Bar Association's Committee on the Federal Judiciary (Chase, 1972; Grossman, 1965), but one can see it elsewhere when newspapers unquestioningly adopt bar association evaluations as their editorial positions on judicial elections.

Court reformers' present consensus, shared by the organized bar, is that judges of the highest quality will be produced by the "merit system," appointment by the chief executive from a list which lawyers play a large role in developing. Such a system may eliminate some of the worst judicial candidates, but it does not meet the high hopes of its proponents; instead of eliminating politics from judicial selection, the politics of the bar may be substituted for more public partisan politics (Watson and Downing, 1969), with visibility lost as a result.

The attention directed to the "best" system of selecting judges has not been matched by attention focused on how to achieve accountability other than to the bar, and the implications of adopting a lawyer-dominated selection process are relatively ignored. Verbiage about the part elections play in "merit system" plans, when judged against the facts, turns out to be largely rhetoric. "Missouri plan" judges' names appear on retention ballots after their initial appointment and again at the end of each term in office, but the retention ballots are not competitive (Shall Judge X be retained?) and the instances in which judges have not been retained are few and far between. It is also often true that when formal rules dictate that judgeships be filled initially by election, the de facto system is often one of appointment, the retiring judge stepping down to allow the governor to fill the seat, so an opportunity for accountability to the broader community is lost.

A number of factors militate against either initial or later judicial accountability through elections. Use of the nonpartisan ballot hinders telling good candidates from bad, particularly if they have no previous political party identity. The norm that sitting judges should not be opposed often leaves the the voter without a choice, even when judges must seek a new term in a formally competitive election. Competition, at times stiff, can occur, but it is infrequent and occurs with a reduced electorate as well (Jacob, 1966; Ladinsky and Silver, 1967), a function of judges' low visibility unless they have been involved in a long, spectacular or controversial trial. (Ask your students or friends how many local judges they can name.) Indeed, Alabama federal judge Frank Johnson, an exception, was kept in the news at least as much by George Wallace's campaigning for office against his decisions as by the decisions themselves, but in any event Johnson, a lifetime judge, did not face election.

While the organized bar prefers selection by appointment (direct or through some variant of the Missouri plan) or, failing that, nonpartisan election, those active in the broader political arena often wish a broader base for selection and removal so as to provide some political accountability, at least to the political party organization if not to the entire political community. The distinction between the organized bar and political party interests is, of course, not hard and fast, because many lawyers are active in partisan affairs. However, the lawyers engaged in machine politics are not likely to be the same ones devoting their principal attention to the bar association. Even the upper-middle-class lawyers involved in "amateur politics" (Wilson, 1962), closer in socioeconomic characteristics to the bar's leaders, generally are different from them, being far more issue-oriented as well as far more liberal.

If initial selection is by partisan ballot, there is at least the possibility of accountability at that time to a constituency represented by one of the political parties. If selection to higher courts is by partisan ballot, continuing accountability to the party may also occur because judges are forced to pay some attention to the party organization if they wish to be nominated for advancement. Judges not seeking higher judicial office need not feel continuing obligation to those who have selected them and are unlikely to do so. Selection which is in some way dependent on the party organization does seem to produce a judiciary more responsive to a different set of interests, for example, ethnic and racial groups, while other selection plans result in a more white, upper-middle-class orientation paralleling concerns of the organized bar (see Levin, 1977). Political party affiliation of judges also produces some continuing accountability insofar as it seems to affect the direction of judges' decisions on some types of issues (Nagel, 1961; Goldman, 1975).

Until recently, removal of judges from office has received much less attention than their selection. Some states provide for recall of judges, and the 1977 removal of Judge Archie Simonson in Wisconsin after his comments on rape suggests that recall can be used to achieve accountability when a judge seriously offends at least some portion of the public. Yet judges' previously-mentioned low visibility which makes defeating judges at reelection or retention time difficult also makes recall an unlikely tool for achieving accountability unless a judge's action is outrageous—and then it may not accurately represent the judge's long-term competence. Removal of judges by recall or defeat is thus an erratic mechanism. A recent Illinois incident in which the former Chief Criminal Judge in Cook County failed to win retention appears to have turned more on ballot position (first on a long list) and ethnic unhappiness over the Democratic organization's earlier refusal to slate a Polish candidate for another judgeship than on the defeated judge's judicial behavior.

The ineffectiveness of other methods of holding judges accountable to certain standards of competence and behavior has in recent years led to greater attention to collegial methods of discipline. All but a few states, following California's lead, have developed judicial discipline commissions which can receive and investigate complaints, with final action usually taken by the state's highest court. A senile California Supreme Court judge was recently removed through such processes and in the past few years several Illinois judges have been reprimanded, suspended, or removed from office. The difficulty, as well as inappropriateness, of using impeachment to remove federal judges and the lack of other mechanisms for relieving them of their duties (see *Chandler* v. *Judicial Council of the Tenth Circuit,* 1970), has led to congressional consideration of a comparable device at the federal level based on the constitutional provision that judges serve "during good behavior." Because most state judicial discipline commissions are relatively new, to say that they have contributed significantly to judicial accountability would be premature. However, they do promise to be an increasingly important weapon in the arsenal of available tools to establish accountability, although, because they place decision-making in the hands of the judges themselves, they will continue the pattern of allowing a professional group to control its own rather than opening up the process to the participation of others.

POLITICAL ACCOUNTABILITY I: POLITICAL ENVIRONMENT, PUBLIC OPINION, AND STRATEGY

Political accountability is accountability to the broad political environment in which judges find themselves and with which they must deal. One form of such accountability occurs when judges' decisions reflect values of the entire political community. This was noted in our discussion of the values affecting local criminal courts. That judges' decisions to some extent reflect constituency characteristics and values should not be surprising, because both federal and state judges are selected from the area their courts serve. As Richardson and Vines remark, "The location of federal courts throughout the states and regions renders them usually susceptible to local and regional democratic forces" (Richardson and Vines, 1970:10; see also 93-100). The pull of the local constituency, even for appointed judges, can be seen in the desegregation cases following *Brown* v. *Board of Education* (Peltason, 1961). However, judges do not always fully internalize their areas' political values. For example, there is little support for the hypothesis that federal judges most closely tied to the South or to the local community allowed the most school segregation, although judges were firmer in limiting segregation in school districts away from the seat of the court than nearby (Giles and Walker, 1975:925, 931).

Judges' lack of uniformity in reflecting local values may result from variations in their political socialization or from the interplay between that socialization and judicial roles. It can be seen in race relations cases, where some state supreme court judges (States' Righters) emphasized local needs and problems and stressed the predominance of state law and judicial processes, while others (Federals) were more willing to decide cases consonant with the U.S. Supreme Court's more liberal interpretation of the U.S. Constitution (Vines, 1965). Some Southern federal judges, confronted with suits on the enforcement of voting rights legislation, were active, even aggressive enforcers of those rights. Others, more gradualist, enforced the law if given sufficient evidence, which would apparently overcome personal bias; still others were highly resistant (Hamilton, 1973).

Public opinion about the courts, their decisions, and the topics with which the courts must deal constitute another specific element in the political environment which may serve to produce accountability. We have little information on public opinion about the state courts, whose lack of visibility would likely produce many "Don't Know" poll responses. We must thus turn our attention to public opinion about the U.S. Supreme Court, about which we know more. People answer Harris and Gallup Poll questions about their general approval of the Supreme Court's work and whether the Court is in touch with the nation, but their level of knowledge about particular cases is on the whole quite low, with most likes and dislikes about the Court based primarily on certain exceptionally well-known decisions, such as those on school desegregation, school prayer, defendants' rights, and the death penalty (Murphy et al., 1973).

Increased knowledge appears to lead to increased disapproval of specific decisions, although diffuse support for the Court does exist alongside such specific dislikes (see Kessel, 1966; Dolbeare, 1967). Negative opinion is not, however, likely to be effective in producing accountability, because quiescence rather than rebellion characterizes the situation; those unhappy about particular decisions seldom indicate they would do much more than contacting their legislators to complain. (However, as discussed below, the hostility can become so great it results in defiance.) Nor does increased education seem to make people more aware of the Supreme Court's political role. Instead it seems to produce increased belief in myths about the Court's decision-making, for example, that it finds rather than makes the law (Casey, 1974).

Indeed, judges can increase their discretion—and thus decrease their accountability—by manipulating the myth that they only find the law, and frequently restating the myth of judicial independence. Although some observers of the courts will remain skeptical, greater use of rhetoric that decisions reached by the justices are compelled by the Constitution and precedent may quiet at least some criticism of the results the judges reach. The Warren Court's lessened attentiveness to the need to manipulate the myth helped heighten the level of criticism its rulings produced. Had *Brown* v. *Board of Education* sounded less like a Sunday school sermon and been more a long, exhaustive, and perhaps boring analysis of the case law of the previous half-century, showing that *Brown* was only an incremental change from the immediately previous law school cases, at least some critics—from whom other members of the public took their lead—might have been pacified (Wasby et al., 1977:97-98).

Even if opinion about the Court and its rulings is relatively undeveloped, courts do seem to be affected by public opinion on a continuing basis although the public does not know it. There are no regular structural arrangements for the transmission of public opinion to the courts other than as lawyers (infrequently) mention it in their briefs, yet the idea that the Supreme Court "follows the election returns" [read: public opinion polls] is hardly new. The Court does not have to announce that its decisions are based in part on public opinion for public opinion to have an effect, as it must to some degree with any political institution. What this means is that the linkage between public opinion and judicial decisions is "noncoercive," but there can be little doubt that judges are aware of public opinion and do respond to it.

We are not without evidence on this matter. The justices have even at times made explicit references to public opinion although without agreement on what direction, if any, it would lead the Court, as in the death penalty cases (*Furman* v. *Georgia,* 1972; *Gregg* v. *Georgia,* 1976). At most other times we must infer its effect. For example, the Warren Court's refusal to give retroactive effect to most of its criminal procedure rulings after public opinion to those decisions had been negative can be seen as a strategic response to hostile reaction (Fahlund, 1973). Another instance is the Court's 1958-1959 withdrawal from its liberal internal security position of the two previous years, after Congress gave serious consideration to the Jenner-Butler bill which would have stripped the Court of some of its appellate jurisdiction (see Sheldon, 1967); larger majorities in cases when the Court was operating under conditions of threat provides still more evidence (Rohde, 1972).

Strong evidence from the trial court level is also now available. In their sentencing of draft offenders during the Vietnam war, "federal district judges responded to public antipathy toward the war by sentencing less harshly those undermining the war effort by draft resistance" (Cook, 1977:579). The correlation between public opinion and sentences was an amazing *.995,* with public opinion lagged by one year against sentencing, and public opinion explained "almost all the variation in draft sentencing from 1967 to 1975." (Interestingly, jury verdicts had little practical district-by-district effect on judge decisions.) The result was that judges were more responsive to public opinion about the war than were legislators or the President.

Public opinion apparently reinforces "the prebench socialization which involved the internalization of community cultural norms" (Cook, 1977:569). Just as their response to other environmental characteristics varies, judges' reaction to public opinion is not uniform, being affected by their orientations and courts' structural characteristics. Judges who are "strict legalists" are less likely to allow themselves consciously to be affected by such external influences. One might also speculate that, although it is considered legitimate for judges to consider the content of *amicus curiae* presentations, "strict legalists" would be less likely to pay heed to the number of such presentations or the identity of the organizations involved, perhaps even giving less weight to the presence of the U.S. government, which tends to fare particularly well when it makes an appearance in cases.

If public opinion serves to produce increased accountability when the direction of that opinion is clear and strong, there are also times when the public—or rather the relevant public*s*—is fragmented or confused, thus providing courts much room in which to maneuver. Similarly, where public expectations differ from the basic expectations held by the bar—when the public's view that the courts "do justice" or its current view of what "justice" entails is at variance with lawyers' perspectives, judges' discretion is increased as they have a choice as to which of their potential audiences they will address their decisions. For example, the public's concern that society's needs are being ignored by attention to criminal defendants' rights may allow a court to skirt attorneys' demands that new procedural protections be enforced. If the public is more conservative in that situation, in others it may be less so. Thus courts may have been able to set aside the organized bar's unwillingness to expand lawyer advertising because of public criticism of the organized bar and greater public demand for information about legal services.

The examples cited above make clear that at least some judicial actions are a matter of strategy. There is not likely to be a "strategy of the court" which all the judges or even a majority have planned collectively. This does not, however, mean that individual judges do not engage in conscious consideration of strategy, with the Court as a whole moving strategically because of values and considerations the justices share (Murphy, 1964; Wasby et al., 1977). To say that courts engage in strategy might seem to indicate that they control the situations about which they plan, but one can instead look at strategic moves as further evidence of judicial accountability to the political environment; if judges completely dominated their environment, they could act with little if any regard to it, knowing their orders would be obeyed almost regardless of how they ruled.

Sometimes a court is quite explicit in indicating awareness of the potential effect of its decisions, true of the Supreme Court in its criminal procedure retroactivity decisions, where the effect on the administration of justice actually became one of the criteria for determining whether to make a new criminal procedure rule retroactive. However, that a court is most effective when it acts most like what the public believes a court should be, thus reinforcing myths, means that the Court will seldom openly address questions of strategy.

Depending on circumstances, considerations of strategy can lead to either activism or restraint. If judges have worried about negative effects of a ruling on a court's life as an institution, as Justice Frankfurter did in his prophecy-of-doom dissent in *Baker* v. *Carr* (1962), and the predicted disaster does not occur, the Court may be encouraged to increase its activity, as it did with reapportionment. Where a court has been "burned," it can also avoid further activity, seen in the almost total absence of school desegregation decisions from *Brown II* (1955) until 1968 (Wasby et al., 1977). While the Burger Court's movement toward a more conservative stance on civil liberties matters may have resulted simply from the substitution of four Nixon appointees for more liberal justices, it is also likely that it reflected the justices' strategy-based view that the Court's position should be more consonant with increasingly conservative public opinion on civil liberties matters to avoid attacks of the sort the Warren Court encountered.

Judges' strategic consideration of their environment is one reason, along with their prior socialization and the processes by which they are selected, why the Supreme Court has been said to have been seldom out of line with the nation's dominant political interests (Dahl, 1957), another way of saying it has been politically accountable, at least in the sense of operating within consensual bounds. In making this argument, Dahl said the Court could take action within relatively narrow limits when other political actors were unable to decide, but could do little without support from Congress and the President. Dahl's further argument that the Court conferred legitimacy on the dominant coalitions' policies has been successfully attacked by a demonstration that he overestimated the Court's ability to legitimize (Adamany, 1973). The Court's accountability, at least to elites, is brought into focus if, instead of the Court's conferring legitimacy on elite policy, elites serve to legitimize the Court. On the other hand, judicial accountability seems reduced if we accept the well-supported argument that Dahl *under*estimated the Court's effectiveness (Casper, 1976). The Court's not infrequent invalidation and delay of policy enacted by the other branches does not, however, indicate the Court's lack of accountability. If the other branches do not override the Court despite an *opportunity* to do so, their failure to act may indicate implicitly that other government officials are not sufficiently exercised to act; even if they would prefer a somewhat different result, they seem to accept the Court's actions as having taken place within acceptable bounds. If courts are responsive to public opinion despite the lack of formal links between that opinion and the courts, thus producing considerable political accountability, heightened accountability might occur if people understood the Court's role as a political institution; if they realized that courts paid attention to public opinion, they might make more instrumental use of their views of the Court's work.

POLITICAL ACCOUNTABILITY II: RESISTANCE

Perhaps the best indicator of judicial accountability is that compliance with courts' rulings is not always immediate or complete. Not only are rulings appealed, with reversal serving to produce accountability, more important, they are ignored, resisted, attacked, and overturned in other arenas. Those who would uphold the idea of "a government of laws" may view with distaste others' refusal to accept judicial rulings, but such action is certainly an often effective means of holding the courts accountable to what important segments of the public view as politically proper.

Because we expect prompt compliance, noncompliance may have received disproportionate attention and routine obedience insufficient attention (Baum, 1978). Yet reversing courts, even the Supreme Court is well within the realm of possibility and noncompliance does occur, abundantly evident with school desegregation, where we found programs of Massive Resistance, violent responses to federal judges' desegregation orders, and the absence of even token desegregation in Deep South states for over 10 years after the Court ruled. Disobedience of the school prayer rulings, while varying from one section of the nation to another (Way, 1968), was at least as great, if somewhat less visible because it was hidden behind school principals' statements that they had told teachers no prayers should be said (Dolbeare and Hammond, 1971).

What actions do the other branches of government take with disliked judicial rulings? Again we can draw examples from the Supreme Court's decisions, for which we have a more complete picture (Wasby, 1970) than we do for other courts. Even the most difficult way to reverse the Court, by constitutional amendment, has occurred several times, not only to change our basic definition of citizenship and to provide for the rights of blacks (the 13th, 14th, and 15th Amendments), but to allow a federal income tax and to give 18-year-olds the right to vote in state and local elections (the 16th and 26th Amendments). Reversals of the Court's rulings interpreting congressional acts are far more common: Congress revised the Court's actions 50 times between 1944 and 1960 (Krislov, 1965:143), in some instances returning the law to its original form, in others altering it to embody what Congress felt was the original interpretation the Court had misread. Action to remove some of the Court's jurisdiction is attempted only infrequently and has succeeded only once, but such attempts can create an atmosphere of threat to which the Court does respond (Rohde, 1972). In addition, Congress has conveyed its views by denying judges pay increases and by intensively questioning judicial nominees, as the Senate Judiciary Committee did when Justice Abe Fortas was nominated to be Chief Justice.

Reaction in local communities is less clear, partly because of uncertainty as to what constitutes compliance. If government officials change an invalidated statute or regulation only modestly or take the minimal action required by a judicial ruling but act counter to its spirit, are they in (technical) compliance or are they evading the Court's ruling? (Wasby, 1970:30-32) Another part of the problem of noncompliance or limited impact is ignorance. What may at first appear to be less than full, willing compliance may be the result of lack of knowledge of judicial rulings. For example, the police generally know that improperly seized evidence will be excluded from trials, but they may not know what constitutes a proper search because no one has taken the responsibility to train them in the intricacies of search law (Wasby, 1976). Ignorance of the law may be no excuse for a criminal defendant but local officials should not be fully blamed for the absence of effective transmission lines to them from the courts.

When noncompliance is clear and does not stem from ignorance, many factors explain its existence. Some, like the ambiguity of the judges' opinions, are case rather than audience characteristics. Follow-up enforcement by government officials is another factor of some importance. Still another, which bears directly on accountability, is the public's feeling as to whether the courts are acting legitimately (Wasby, 1970:29, 265). The courts have shared in the over-all loss of credibility recently suffered by all government institutions, but the courts cause particular problems for themselves when those who pay closest attention come to believe that they have mishandled problems because they "went too far" or lacked the necessary knowledge of the "real world" to be affected by their rulings. Such beliefs can easily lead to resistance, which is one way to make an institution thought not to be acting in legitimate ways alter its behavior; this is the case even if the public engages in resistance because it does not know how else to cope with intensely disliked rulings, as its earlier-noted failure to follow through on its opinions about cases shows.

When such situations develop, how can we deal with them? The courts cannot back down too readily or openly in the face of intransigence, for to do so would disappoint other elements of their constituencies and tarnish the myth that courts are apolitical discoverers of the law, not its makers. If judges were more fully accountable, dislike might not take the forms it now takes from time to time, yet judges are intended to be disliked: that is why we try to give them some basis for effective independence. However, if they become too estranged from the public, they will not be allowed to exercise their independent judgment in the short run. If we do not want public resistance to undermine judges' authority, we must do more than we do at present within the adversary system to inform the courts about the environment they affect (Wasby, 1978b), in the hopes that they will be less likely to offend affected populations. At the same time, more must be done to educate the public about the operation of our judicial system. Whether we need as well to develop alternative mechanisms of judicial accountability is not clear. Perhaps better use of present mechanisms will be sufficient, but we need to make better use of them before we can be sure.

A CONCLUDING NOTE

Where does all this leave us? With a mixed picture, just what we would expect to find with respect to the accountability of *any* political institution. The posture of critics that courts usurp authority properly belonging to elected officials, coupled with the charge that judicial review is undemocratic, may make it surprising that we find any accountability of the courts. Part of the reason is that social scientists have seldom looked at courts from this perspective, although reformers have often tried to achieve it as they see it—without a realistic picture of the judicial process. Our exploration indicates that there are a variety of both formal mechanisms and general political responses by which judges can be held to account and that, perhaps more often than we would presume, the courts are in fact so held from time to time.

Do the mechanisms of accountability identified here push or pull in any single direction? Our tentative answer is that, as identified here, accountability within the legal system and political accountability seem to move in opposite directions. Within the legal system judges' accountability is mostly to lawyers or to other judges, through socialization to the legal profession, precedent, reversal, the norm of judicial independence itself, and organizational needs. On the other hand, political accountability pulls judges toward the broader public—the entire political community—or at least its more attentive members; this occurs particularly through judges' attention to their environment including public opinion, their use of strategy, and by resistance to judicial rulings. The direction in which judicial selection pulls depends on which mechanism is used, one like merit selection, in which lawyers dominate, or partisan election, which produces responsiveness to a broader set of interests.

There is, however, some overlap and possible convergence between the two broad types of accountability. This occurs because lawyers simultaneously play a large role in judicial selection and are members of the political elite. Lawyers' professional associations are also not substantially out of line with broader public views, as we can see from the American Bar Association's gradual shift from a markedly conservative ideology to a far more moderate posture both on matters directly affecting the legal profession like delivery of legal services and on more directly political items like the decriminalization of marijuana. Perhaps more important, the values and attitudes of the elite constitute a large part of judges' environment. When appellate judges developing doctrine for lower court judges base their rulings on the environment, they draw on the political and legal culture from which the lower court judges will have absorbed values before joining the bench and from which they continue to draw. In particular instances or in the short run, political accountability and accountability within the legal system may diverge, running in different directions. However, over the longer haul, important conjoint legal and political system constraints can operate to limit judges' discretion and thus to hold them accountable to those they are supposed to serve.

REFERENCES

ADAMANY, D. (1973). "Legitimacy, realigning elections, and the Supreme Court." Wisconsin Law Review, 1973, 3:790-846.

BAAR, C. and BAAR, E. (1977). "Introduction: Judges and court reform." *Justice System Journal,* 3 (Winter):99-104.

Baker v. Carr, 369 U.S. 186 (1962)

BALBUS, I. D. (1973). The dialectics of legal repression: Black rebels before the American criminal courts. New York: Russell Sage.

BAUM, L. (1978). "Lower-court response to Supreme Court policies: Reconsidering a negative picture." Justice System Journal (forthcoming).

--- (1976). "Implementation of judicial decisions: An organizational analysis." American Politics Quarterly, 4 (January):86-114.

BEISER, E. (1972). "Lawyers judge the Warren court." Law & Society Review, 7 (Fall):139-149.

BERGER, R. (1977). Government by judiciary: The transformation of the Fourteenth Amendment. Cambridge, Mass.: Harvard University Press.

BERMAN, H., and GREINER, W. (1972). The nature and functions of law (3rd ed.). Mineola, N.Y.: Foundation Press.

Brown v. Board of Education, 347 U.S. 483 (1954) and 349 U.S. 494 (1955)

CANON, B. (1974). "Organizational conumacy in the transmission of judicial policies: The *Mapp, Escobedo, Miranda,* and *Gault* cases." Villanova Law Review, 20 (November):50-79.

--- (1973). "Reactions of state supreme courts to a U.S. Supreme Court civil liberties decision." Law & Society Review, 8 (Fall):109-134.

CARP, R. (1972). "The scope and function of intra-circuit judicial communications: A case study of the Eighth Circuit." Law & Society Review, 6 (February): 406-426.

--- and WHEELER, R. (1973). "Sink or swim: The socialization of a federal district judge," Journal of Public Law, 21, 2:359-393.

CASEY, G. (1974). "The Supreme Court and myth: An empirical investigation." Law & Society Review, 8 (Spring):385-424.

CASPER, J. (1976). "The Supreme Court and national policy making." American

Political Science Review, 70 (March):50-63.
Chandler v. Judicial Council of the Tenth Circuit, 398 U.S. 74 (1970)
CHASE, H. (1972). Federal judges: The appointing process. Minneapolis: University of Minnesota Press.
COLE, G. (1970). "The decision to prosecute." Law & Society Review, 4 (February):313-343.
COOK, B. B. (1977). "Public opinion and federal judicial policy." American Journal of Political Science, 21 (August):567-600.
--- (1973). "Sentencing behavior of federal judges: Draft cases–1972." University of Cincinnati Law Review, 42 (4):597-634.
--- (1971). "The socialization of new federal judges: Impact on district court business." Washington University Law Quarterly (Spring):253-279.
Criminal justice in extremis: Administration of justice during the April 1960 Chicago disorder (1969). Chicago: American Bar Foundation.
DAHL, R. (1957). "Decision-making in a democracy: The rule of the Supreme Court as a national policy-maker." Journal of Public Law, 6 (Fall):279-295.
DIXON, R. G., JR. (1976). "The 'new' substantive due process and the democratic ethic: A prologomenon." Brigham Young University Law Review (1):43-88.
DOLBEARE, K. (1967). "The public views the Supreme Court." Pp. 194-212 in H. Jacob (ed.), Law, politics, and the federal courts. Boston: Little, Brown.
--- and HAMMOND, P. (1971). The school prayer decisions: From court policy to local practice. Chicago: University of Chicago Press.
EISENSTEIN, J. and JACOB, H. (1977). Felony justice: An organizational analysis of the criminal courts. Boston: Little, Brown.
FAHLUND, G. (1973). "Retroactivity and the Warren court: The strategy of a revolution." Journal of Politics, 35 (August): 570-593.
FRANKEL, M. (1973). Criminal sentences: Law without order. New York: Hill & Wang.
Furman v. Georgia, 408 U.S. 238 (1972)
GILES, M. and WALKER, T. (1975). "Judicial policy-making and southern school segregation." Journal of Politics, 37 (November): 917-36.
GOLDMAN, S. (1975). "Voting behavior on the United States courts of appeals revisited." American Political Science Review, 69 (June):491-506.
GRAGLIA, L. (1976). Disaster by decree: The Supreme Court decisions on race and the schools. Ithaca, N.Y.: Cornell University Press.
Gregg v. Georgia, 96 S.Ct. 2909 (1976)
GROSSMAN, J. (1965). Lawyers and judges: The ABA and the politics of judicial selection. New York: John Wiley.
HAMILTON, C. (1973). The bench and the ballot: Southern federal judges and black voters. New York: Oxford University Press.
HOROWITZ, D. (1977). The courts and social policy. Washington, D.C.: The Brookings Institution.
JACOB, H. (1966). "Judicial insulation–elections, direct participation, and public attention to the courts in Wisconsin." Wisconsin Law Review (Summer): 801-819.
KALVEN, H. and H. ZEISEL (1966). The American Jury. Boston: Little, Brown.
KESSEL, J. (1966). "Public perceptions of the Supreme Court." Midwest Journal of Political Science, 10 (May):167-191.
KRISLOV, S. (1965). The Supreme Court in the political process. New York: Macmillan.
LADINSKY, J. and SILVER, A. (1967). "Popular democracy and judicial independence: Electorate and elite reactions to two Wisconsin Supreme Court elections." Wisconsin Law Review, 1967 (Winter):128-169.

LEVIN, M. (1977). Urban politics and the criminal courts. Chicago: University of Chicago Press.

Morgan v. Hennigan, 379 F.Supp. 410 (D.Mass. 1974), aff'd sub nom. Morgan v. Kerrigan, 509 F.2d 580 (1st Cir. 1974), cert. denied, 421 U.S. 963 (1975), 401 F.Supp. 216 (D.Mass. 1975), aff'd, 530 F.2d 401 (1st Cir. 1976), cert. denied sub nom. McDonough v. Morgan, 96 S.Ct. 2646 (1976).

MURPHY, W. (1964). The elements of judicial strategy. Chicago: University of Chicago Press.

--- (1959). "Lower court checks on Supreme Court power." American Political Science Review, 53 (December):1017-1031.

---, TANENHAUS, J., and KASTNER, D. (1973). "Public evaluation of constitutional courts: Alternative explanations." Sage Professional Papers #01-045. Beverly Hills, Cal.: Sage Publications.

NAGEL, S. (1961). "Political party affiliation and judges' decision." American Political Science Review, 55 (December):843-851.

Nebraska Press Association v. Stuart, 96 S.Ct. 2791 (1976)

PELTASON, J. (1961). Fifty-eight lonely men. New York: Harcourt, Brace, & World.

RICHARDSON, R. and VINES, K. (1970). The politics of federal courts. Boston: Little, Brown.

ROHDE, D. (1972). "Policy goals and opinion coalitions in the Supreme Court." Midwest Journal of Political Science, 16 (May):208-224.

SCHMIDHAUSER, J. and GOLD, D. (1962). "*Stare decisis,* dissent, and the background of justices of the Supreme Court of the United States." University of Toronto Law Journal, 14 (May):194-212.

SHELDON, C. (1967). "Public opinion and high courts: Communist party cases in four constitutional systems." Western Political Quarterly, 20 (June):341-360.

SIMON, R. J. (1967). The jury and the defense of insanity. Boston: Little, Brown.

United States v. Butler, 297 U.S. 1 (1936)

VINES, K. (1965). "Southern state supreme courts and race relations." Western Political Quarterly, 13 (March):5-18.

WASBY, S. L. (1978a). "Communication within the Ninth Circuit Court of Appeals: The view from the bench." Golden Gate University Law Review (forthcoming).

--- (1978b). "Judicial capacity to resolve social problems: A law brief and a social science response." Vanderbilt Law Review (forthcoming).

--- (1976). Small town police and the Supreme Court: Hearing the word. Lexington, Mass.: Lexington Books.

--- (1970). The impact of the United States Supreme Court: Some perspectives. Homewood, Ill.: Dorsey Press.

---, D'AMATO, A., and METRAILER, R. (1977). Desegregation from Brown to Alexander: An exploration of Supreme Court strategies. Carbondale, Ill.: Southern Illinois University Press.

WATSON, R. and DOWNING, R. (1969). The politics of the bench and bar. New York: John Wiley.

WAY, H. F. (1968). "Survey research on judicial decision: The prayer and Bible reading cases." Western Political Quarterly, 21 (June):189-205.

WILSON, J. Q. (1962). The amateur Democrat: Club politics in three cities. Chicago: University of Chicago Press.

Wyatt v. Stickney, 325 F.Supp. 781 (M.D.Ala. 1971), 334 F.Supp. 1341 (1971), 344 F. Supp. 373 (1972), 344 F.Supp. 387 (1972), aff'd in part sub nom. Wyatt v. Aderholt, 503 F.2d 1305 (5th Cir. 1974).

7

Accountability in Policy Process: An Alternative Perspective

VIRGINIA GRAY

☐ IN DEMOCRATIC GOVERNMENTS the public through the election of representatives and the creation of a bureaucracy gives its officials independence in enacting and implementing public policy. Yet the public retains ultimate control through the threat of removal from office. While there is always a tension between independence and control, in the late 1970s many perceived too much independence on the part of elected and appointed officials and thus a need for more accountability or control.

Other chapters in this volume examine the performance of institutions (the legislature, judiciary, bureaucracy) in an effort to assess their accountability to the public. The institutional focus tends to adopt a formalistic definition of accountability which, as Pitkin (1967:55) warns us, ignores the substance of the actions taken by the representative. For this reason she dismisses accountability as a minor strand of representation theory. The institutional focus also tends to ignore what I see as the core of the current public dissatisfaction with government: disappointment with the outcome of public policies.

The public is demanding substantive accountability or control over the results of governmental programs. This trend was especially evident in the 1976 elections when the nationwide popularity of Governor Jerry Brown of California became apparent, and other neo-conservative politicians succeeded at the polls as well. In fact, one of Brown's first acts as governor demonstrates this concern with accountability for results. He made California's continuation in the Safe Streets program (a program in which federal dollars are given to state governments on a block grant basis to be spent on criminal justice projects of the state's choosing) contingent upon a reduction in crime by the end of a one year period. In Governor Brown's opinion (Advisory Commission on Intergovernmental Relations, 1977:275) citizens are most concerned with crime control and any attempt to justify the expenditure of $190 million on system improvement or other grounds is false and misleading.

The best guarantee of control is designing accountability systems which are appropriate to the policy, rather than specific to the institution. It seems to me that substantive policy areas vary in the degree to which their outcomes can be assessed. Likewise, the policy makers in each stage of the policy process vary in the degree to which they account to the public. My task is to discuss how accountability might be facilitated throughout the policy process.

At an earlier conference (Smith, 1971:29) three types of accountability were identified: fiscal, process, and program. Fiscal accountability is the easiest to achieve: was the money spent in accord with proper accounting practices for the purposes authorized by law? The outrage over campaign finance irregularities suggests that the public demands at least fiscal accountability. Process accountability asks: did the agency implement the policy in a way that should produce results? Were schools built, teachers hired, teaching materials purchased, and pupils in attendance the required number of days? Measures of effort serve as proxies for the quality of the result. More recently, process accountability has come to include whether the process is open to blacks, women, minority groups, the aged, and the handicapped. The type of accountability most difficult to achieve and to assess is program accountability: did the policy produce the intended results? Policy impact measures are required here.

The types of accountability represent an evolution in the public's demand for accountability. In the late 19th century the public began to demand that their elected officials distribute offices on the basis of merit instead of politics. In more recent times process requirements were added: competitive bidding for materials, affirmative action in hiring, planning, intergovernmental approval, earmarking of funds for specific purposes within a broad grant of authority. The current interest in accountability seems to focus on the results of programs and would seem to require systematic evaluation.

It is important to know which of the three types of accountability is appropriate for various kinds of activities. When the problem is new and solutions unknown, fiscal accountability may be the most one can ask for. Federal basic research programs have relied on this argument for years. This justification for existence in turn leads Senator William Proxmire and others to hold researchers to account for their fiscal procedures and time allocations.

For some engineering fields program accountability is possible. Production may be shut down if a weapon does not perform as promised. In some cases the citizen can conduct his own program assessment. A recent national survey for the U.S. Senate (Chelimsky, 1977:8) found that the only government worker to get high marks from the public was the local trash collector, because at least people knew whether he was doing his job or not.

Traditionally, in social programs only process accountability has been possible, but in the past few years some impact evaluations have been performed. Especially since the demise of the Great Society, the public's demand for results seems to focus on social programs. Recently President Jimmy Carter's reorganization team asked Congressmen to tell them which departments their constituents found most frustrating and confusing. A compilation (Washington Post, 1977:A3) of the replies from 200 offices showed that the bulk of the complaints centered on four departments: Justice, Labor, Housing and Urban Development, and Health, Education and Welfare. Thus the social areas draw the most fire.

While evaluation is the key to accountability, we also need to look at the other stages of the policy process because they contribute to the policy outcome and they offer different opportunities for accountability. The first phase of the policy process is problem identification and agenda formation. Bureaucrats operating a program, as well as a President or congressman hearing from constituents, all may define a situation as a problem, but only a congressman can actually introduce a bill and thus place an issue on the formal agenda. They are held accountable for their problem recognition. For example, the President and Congress are presently being held to account for not recognizing our nation's energy problems sooner.

The speed with which issues go from the informal public domain to the formal agenda of an institution such as a legislature or court varies according to a number of factors enumerated by Cobb and Elder (1972), the most important citizen strategy being issue expansion (an expansion in the number of participants). In a study of agenda setting in the U.S. Senate, Walker (1977:428) found that legislative activists tended to be young, liberal Senators from the larger, more politically competitive states. Thus party competition seems to be a condition which encourages representatives to introduce more legislation, an activity for which they are accountable.

Likewise, the mass media, interest groups, and researchers are frequently involved in perceiving a situation and defining it as a problem which needs governmental attention. Part of the current accountability movement is the vague but frequently expressed notion that all major social institutions have some sort of obligation to account to the public. The mass media is held responsible particularly for its actions or inactions in agenda formation. The latest national survey of confidence in major institutions (Washington Post,

1978:A8) shows a slight decline in the public's confidence in the press, whereas confidence has risen dramatically since last year for most institutions. Though it is often difficult to discern whether newspapers report the agenda or structure the agenda, Walker (1977:442) concludes that the *New York Times* at least follows the lead of Congress. In each of the three cases of safety legislation he studied, the newspaper began extensive coverage of the issue in the year it came to the floor of Congress and then maintained an interest in the issue for a few years afterward while the new legislation was being implemented.

Next is the policy formulation stage in which there is a proposal to resolve the problem. Virtually all of the actors from the previous stage except the media are involved in formulating alternative proposals. In most cases the primary actors are likely to be the executive departments (bureaucrats and political appointees) and Congress. This is the stage where I think it is hardest to identify accountability. In the problem definition stage it is relatively easy to detect a broad public sentiment that something be done about crime, for instance, and to see if that sentiment is translated into a formal proposal to be considered by an institution. But it is difficult to tell if a specific proposal or even set of alternatives represents what the public wants done, primarily because the public often only wants "something done."

The public might reasonably expect to hold formulators responsible for proposing a policy which, given some broad public sentiment to do something about the crime problem, has a reasonable chance of doing something about that problem. That is, the public might expect that formulators be responsible for their policy theory and that President and Congress remove staff and bureaucrats whose theory has been shown to be wrong.

To give an example, according to Moynihan (1973) at least, in designing the Family Assistance Program, his group discovered that the prevailing theory about the relation of unemployment and the welfare rolls was simply wrong. Therefore, the previous group of policy formulators might have been responsible for designing a policy doomed to failure. To some extent the demise of social workers and the rise of economists in the formulation of welfare policy may be due to the fact that the social work profession was held to account by Congress and President Nixon for its theory of welfare reform and it was found lacking. But though there are several examples of the removal of advisors who were wrong on some issue, more often it is not possible to trace the lack of expected outcome to the theory.

One condition which would seem to inhibit accountability is formulation by subgovernments (governmental and nongovernmental experts who make most of the routine decisions in a substantive area). As Ripley and Franklin (1976) have shown, subgovernments are most influential on the least visible policies—distributive policies. Furthermore, there is a pressure for regulatory policy (especially the enforcement of) and redistributive policy (e.g., federal aid to education) to be reformulated as distributive policy in order to be passed by Congress. Thus accountability in formulation is more likely when policies remain regulatory or redistributive because the conflicts are more visible and participation broader.

The third stage is the policy adoption (legitimation) stage, dominated by the legislature but including lobbyists, bureaucrats, and others. In this stage it is much easier for the citizen interested in a particular policy area to recognize who is responsible for the passage or rejection of a bill he is interested in. At the polls he can hold to account his representative for his vote on a bill and through interest group efforts he can try to hold to account a committee for its treatment of a bill. The other important actor, by virtue of his veto as well as his leadership on the floor, is the President or other executive.

The flaw in this traditional approach to accountability is, of course, that the American people have not been noted for voting on the basis of ideology, issues, or policies. However, there is evidence that issue voting is on the rise, though the magnitude of that increase and the reasons for it are still being debated. Miller and Levitin (1976:46) argue that contemporary political leaders have increased public awareness of political issues and contributed to a dramatic increase in issue connectedness, a necessary condition for issue voting.

Next is the implementation stage (putting the program into effect) in which the bureaucracy has the prime responsibility. In addition, congressmen doing casework influence how laws are administered. For some voters a congressman's casework may be more important than how he votes on the floor. Also the courts are involved in implementation as they secure compliance with policies promulgated by the legislature.

The bureaucracy is responsive to several pressures. First, the bureaucracy is ultimately responsible to the Congress, which checks on implementation through casework and more systematic oversight activities. Several recent studies (Pressman and Wildavsky, 1973; Van Meter and Van Horn, 1975; see also Sabatier and Mazmanian, 1977) have identified factors, manipulable by Congress, which might foster the ability of citizens to hold bureaucrats responsible: clarity of statutory language, few veto points, adequate money, a new or friendly established implementing agency. Second, bureaucrats who are also professionals respond to their profession's norms and values which may or may not be shared by the public. Third, an agency's need to build a constituency among interest groups and attentive publics constrains it to be accountable to the general public.

At last we get to the evaluation stage where everyone gets into the act of judging the worth of governmental programs. Congress increasingly evaluates agencies through oversight hearings and the requirement of periodic reporting by agencies. Congress is also considering a sunset bill which would require agencies to justify their existence every five years. Agencies employ persons and contract with others to evaluate their programs systematically. The General Accounting Office conducts performance assessments while the Office of Management and Budget required zero based budgeting for FY79. Commissions, study groups, task forces, and private researchers conduct their own evaluations. Interest groups and the media publish ratings and other assessments of programs.

Evaluation, whether systematic or anecdotal, governmental or nongovernmental, is essential to accountability since it gives substance to the formality. Otherwise, there is no information with which to hold bureaucrats and legislators responsible for their actions. Besides the accountability movement's demand for evaluation, it is also probably the case that evaluation (and hence the potential for accountability) has increased because we now have the research technology (e.g., cost benefit analysis) to perform systematic evaluations.

Evaluative activities have grown in all branches of government. From 1969 to 1974 nondefense evaluation spending by federal agencies rose from $20 million to $146 million, an increase of 600%. By 1976 the figure was $240 million according to Chelimsky (1977:1). Likewise, congressional oversight activities are increasing. Aberbach (1977:20) reports that in the House the days of hearings devoted to oversight (as a percentage of all hearing days) increased from 11.2% in the 91st Congress to 18.0% in the 94th Congress and from 11.3% to 18.7% for the Senate.

In both branches of government the problem from an accountability perspective is how to create incentives for public officials to engage in evaluative activities. A recent conference report on the use of evaluation in federal agencies (Chelimsky, 1977) suggests the nature of the incentive problem in the agency context. Three perspectives on evaluation emerged from the three-day conference: (1) knowledge: a rigorous scientific evaluation to provide evidence leading to new knowledge about social problems and about the governmental strategies for addressing them; (2) management: a support system for assessing and improving the operational efficiency of governmental programs where the basic question is "how can we make the program better," rather than "shall we keep it?"; (3) accountability: to make the best possible use of tax resources by holding program managers accountable for the worth (effectiveness and operational quality) of their programs.

The knowledge perspective was preferred by five agencies, the management perspective by two agencies, and the accountability perspective by no agencies. The accountability perspective was, however, preferred by Congress, a few researchers, and some General Accounting Office and Office of Management and Budget representatives present at the conference. Generally, the agency representatives were opposed to any evaluations whose results were publicly available. They felt the results of an evaluation should be known only to the program manager, who would then have the discretion to act upon the findings.

Thus evaluation is viewed as a threat to the program manager and it will take very great positive incentives to overcome this negative aspect. As one participant put it (Chelimsky, 1977:38): "You will find greater interest in Federal agencies in obtaining and spending larger budgets than in getting results." To overcome this obstacle Congress might direct the Office of Management and Budget or the Civil Service Commission or even individual agencies to reward and punish managers for evaluation activities, i.e., to promote successful evaluators and demote managers who refuse to evaluate.

Other deterrents to the utilization of evaluation results mentioned were the timeliness of the evaluation report and the often imprecise Congressional language which makes it difficult to know which goals to evaluate. It is of particular interest to note that while the researchers present thought that methodological problems were the greatest deterrent to the use of evaluation results, the agency participants talked only of political problems and did not think that substantial methodological improvements would lead to greater utilization.

In the congressional context it is equally clear that there must be incentives for individual congressmen to engage in oversight activities. Just as the agency's reward is a bigger budget regardless of program impact, the congressman's reward is reelection regardless of program impact. External incentives for congressional review of executive branch decisions include the following factors: (1) an environment of scarce resources such as our current economic situation. Aberbach (1977:10) suggests that since hard choices must be made, congressmen will find it easier to transfer the cost of making an unpopular decision to the outcome of an evaluation. (2) The new breed of neo-conservative and skeptical congressmen elected by the increasingly conservative American public. *Congressional Quarterly* (1978:116) reports that the 1976 House Democratic freshmen voted more conservatively than other groups (41% of the time). Similarly, a recent poll (*New York Times,* 1978:30) shows that 42% of the public describe themselves as moderately or very conservative and only 23% label themselves liberal. (3) Split partisan control between Congress and the Presidency, and (4) a severe crisis such as Watergate.

Internally there are new structural incentives or opportunities for congressional oversight due to the establishment of oversight committees and subcommittees, larger committee staffs, the new (1974) budget process, and periodic rather than permanent authorization of programs. Aberbach (1977:12) identifies the new budget process as having the greatest potential for increasing oversight activity because authorization committees may compete with each other to present well-documented recommendations, based on evaluations, to the Budget Committee.

In this brief look at each stage of the policy process we have seen that achieving accountability is much more complex than simply measuring the performance of one institution, because most institutions participate in several stages of the policy process and because the substance of the policy determines the type of accountability which can be expected. We have identified in each stage some conditions which may foster accountability: issue expansion and party competition in the agenda formation stage; lack of subgovernments in formulation; issue voting for the legitimation stage; certain organizational and resource conditions for the implementation stage; internal opportunities and external incentives for evaluation and oversight.

Most of these factors are likely to increase given the present political environment. *Newsweek* labels this movement "the new pragmatism" or "the new synthesis" and says (1977b:34) "it is an effort to draw the line on public spending, to apply a stringent cost-benefit analysis to social programs and in general to lower expectations."

Finally we must identify some problems inherent in achieving accountability, whether our focus is on the institution or the policy process. One problem in achieving accountability is that accountability theory assumes that all officials fear removal from office; i.e., it assumes they are politically ambitious. Accountability is impossible unless elected officials desire to remain in office or seek higher office. The high rate of turnover of state legislators (sometimes around half) and city councilmen should make us expect that accountability will be greater at the national level than at the state and local levels. Among states we would expect more accountability where the opportunity structure offers more chances for advancement.

A second problem is the possible tension between process and program accountability. Agency representatives at the Mitre conference felt torn by the conflicting claims of different types of accountability. One said (Chelimsky, 1977:38):

> Need I tell you that it often seems that there is accountability for everything but results? There is accountability for fidelity to a policy line even when the policy is vague or ill-defined. There is accountability for good public and Congressional relations. There is accountability for spending one's money promptly. There is accountability for assuring compliance with a thousand and one Federal laws and regulations and so on. But to make Government work, we must establish accountability of Government managers for program performance and program results.

I think the current controversy over the Bakke case (*The Regents* of *the University of California* v. *Allan Bakke*) illustrates the conflict from the citizen's perspective. According to *Newsweek* (1977a:52), nearly 60 *amicus curiae* briefs have been filed, the greatest number in the history of the Court. If one is interested in program accountability (as I have argued that most are), then an increase in the number of black medical students would be the sole criterion of program success. Thus Cal Davis' plan of allotting 16 places to disadvantaged applicants is a successful affirmative action program. In fact, without affirmative-action programs which applied differential standards to black and white applicants, there probably would be no increase nationally in the number of black medical school students.

However, the separate admissions pools give the appearance of a quota, and many people believe that quotas are unfair and violate some fundamental notion of equality. Though the desired outcome has been achieved, the process is no longer fair. In this line of thinking an affirmative-action program should have goals for black admissions, and medical schools should be judged by their recruiting efforts. As long as the process is fair, the outcome is unimportant.

The tension between the two types of accountability is the tension between conflicting goals—the merit principle and equality for previously excluded (minority) groups. We will always have some conflicting goals, and the argument over which kind of evaluative criteria to use is a reflection of that fact, just as we will disagree over the proper balance between independence and control. In fact, we should begin to plan a sequel to this book, a volume on how to achieve independence in an urban society.

REFERENCES

ABERBACH, J. D. (1977). "The development of oversight in the United States Congress: Concepts and analysis." Paper presented at American Political Science Association meeting, Washington, D.C., September.

Advisory Commission on Intergovernmental Relations (1977). Safe streets reconsidered: The block grant experience 1968-1975, Part B. Washington, D.C.: U.S. Government Printing Office.

CHELIMSKY, E. (1977). An analysis of the proceedings of a symposium on the use of evaluation by federal agencies. McLean, Va.: Mitre Corporation.

COBB, R. W., and ELDER, C. D. (1972). Participation in American politics: The dynamics of agenda-building. Boston: Allyn & Bacon.

Congressional Quarterly (1978) "Freshman House Democrats' 1977 voting record more conservative than '74 class." Volume 36 (January 21): 116-117.

FOX, H. W., HAMMOND, S. W. and NICHOLSON, J. B. (1977). "Foresight, oversight, and legislative development: A view of congressional policy making." Paper presented at American Political Science Association meeting, Washington, D.C., September.

JONES, C. O. (1977). An introduction to the study of public policy. North Scituate, Mass.: Duxbury Press.

MILLER, W. E., and LEVITIN, T. E. (1976). Leadership and change. Cambridge, Mass.: Winthrop Publishers.

MOYNIHAN, D. P. (1973). The politics of a guaranteed income. New York: Random House.

Newsweek (1977a). "The furor over 'reverse discrimination.' " September 26, pp. 52-58.

--- (1977b). "Is America turning right?" November 7, pp. 34-44.

New York Times (1978). " 'Conservatives' share 'liberal' view." January 23, pp. 1, 30.

PITKIN, H. F. (1967). The concept of representation. Berkeley: University of California Press.

PRESSMAN, J., and WILDAVSKY, A. (1973). Implementation. Berkeley: University of California Press.

RIPLEY, R. B., and FRANKLIN, G. A. (1976). Congress, the bureaucracy, and public policy. Homewood, Ill.: Dorsey Press.

SABATIER, P., and MAZMANIAN, D. (1977). "Toward a more adequate conceptualization of the implementation process–with implications for regulatory policy." Unpublished paper, Pomona College.

SMITH, B.L.R. (1971). "Accountability and independence in the contract state." Pp. 3-69 in B.L.R. Smith and D. C. Hague (eds.), The dilemma of accountability in modern government. New York: St. Martin's Press.

VAN METER, D., and VAN HORN, C. (1975). "The policy implementation process: a conceptual framework." Administration and Society 6 (February): 445-488.

WALKER, J. L. (1977). "Setting the agenda in the U.S. Senate: A theory of problem selection." British Journal of Political Science 7 (December):423-445.

Washington Post (1978). "Confidence in institutions up strongly." January 5, pp. 1, A8.

--- (1977). "Agencies wade through flood of gripes from hill." December 7, p. A3.

Part III

Accountability Controls

Introduction

□ INHERENT IN ANY DISCUSSION of accountability is a concern with the strategies and devices that can be used in order to achieve accountability. The question considers controls which are available to citizens individually and collectively in the pursuit of accountability from public agencies. Although some such strategies have been discussed in previous chapters—professionalism, agency performance reviews, policy outcomes, and so on—the chapters in Part III focus specifically on the strategies and devices available in securing accountability.

In Chapter 8, Stephen H. Linder discusses some of the external controls available for securing accountability over the discretion apparent in administrative agency discretion. Obviously, as Linder points out, the greater the specificity of legislative directives to administrative agencies, the lower the discretion level, and the greater the accountability potential. However, short of a complete listing of agency responsibilities and permitted actions, added legislative specification is not likely to reduce agency discretion. Thus the requirement exists for external controls on agencies administered by the courts and by legislative bodies.

The most discussed device for securing accountability in our political system has been, and will probably continue to be, popular elections of public officials. James H. Kuklinski discusses the potential role of elections in legislative bodies in light of recent research concerning the electoral system. His analysis centers on whether electoral marginality serves as a means for securing legislative responsiveness and accountability.

In order to discuss a different strategy for securing accountability through interest groups, Michael A. Baer advocates distinguishing between accountability and responsiveness. Once this has been done, he notes that accountability may work at cross-purposes with the role of interest groups if interest groups seek to have their policy positions embodied in governmental acts.

Finally, Floyd E. Stoner considers the role of auditors in securing accountability. He points out how, as the scope of fiscal auditing has expanded, so, too, have auditors become an important element of accountability. In the expanded usage of various governmental auditors, their role has become one of reviewing agency performance and conformance with legislative intent. What is very clear is that modern fiscal audits include much more than reviewing an agency's financial statement.

8

Administrative Accountability: Administrative Discretion, Accountability, and External Controls

STEPHEN H. LINDER

□ TO MEET THE GROWING DEMAND for government solutions, administrative agencies have been authorized to expand public funds and to impose duties and obligations on citizens. Within a democratic system, these agencies are accountable to the public, through their elected representatives, for the responsible exercise of this power. Responsibility for delegated power implies the use of discretion and judgment, based upon a weighting of values (Pennock, 1960). Under broad delegations of power, concern focuses upon the scope of administrative discretion: to whose or to what values are administrators responsive?

Accountability refers to the enforcement of responsiveness through procedures that affect the value premises of administrators' decisions. Elections are the most familiar procedures for the accountability of public officials. However, since so few administrators are elected, accountability to the public is promoted by other institutions. Legislatures and courts rely upon formal controls on administrative discretion to enforce responsiveness to certain values. Other measures are available for insuring accountability, including reliance on informal relationships or upon internal controls (Gilbert, 1959); but these are supplementary to the formal controls associated with the separation of powers. This essay discusses legislative and judicial controls as a response to problems that are endemic to administrative discretion. In this context, the need for accountability can be contrasted with the limitations of controls and with recent procedural developments designed to correct these limitations.

DISCRETION AND ACCOUNTABILITY

The delegation of authority has created administrative discretion over the implementation of statutes. Discretion permits interpretation of responsibilities and selection of value premises to underlie decisions. Left to supply their own premises, administrators' choices will reflect the biases of their situation, their survival needs, relations with clientele, and personal inclinations. The quest for accountability is a response to the biases associated with discretion. Formal controls place restrictions on the exercise of discretion, both to compensate for this bias and to stimulate responsiveness to institutional values.

DELEGATION OF AUTHORITY

The imposition of formal, judicial, and legislative controls to achieve administrative accountability must be placed in the context of the debate over administrative discretion. According to one view (Hyneman, 1950), there should be detailed legislative authorizations behind all administrative activity. The legislature is authorized to make choices among values, by the electoral consent of the public. Hence, administrative actions which impose obligations on the public are legitimate, only to the extent that they conform with legislative directives. To the extent that discretion is delegated to the agency to apply directives, their decision procedures must be open to judicial scrutiny. The function of the courts in this scheme is to restrain agency departures from compliance with legislative authorizations; review focuses on *ultra vires* activities, as well as unreasonable interpretations of legislative directives. In this instance, restricted delegations of authority to administrative agencies are not precluded, but extensive discretion is taken as an abuse of popular sovereignty.

From its initial position against any delegations, the Supreme Court was to focus on the issue of standards to circumscribe delegations of power to agencies. Despite the increase in the frequency of broad grants of power, this principle was applied only twice.[1] In both cases, delegations to the National Recovery Administration were invalidated, not for their scope, but because they lacked standards and procedural safeguards (Gellhorn, 1972). Rather than restricting broad delegations, this precedent imposed acceptable conditions upon them. Nonetheless, those who objected to the discretionary exercise of broadly delegated power on the grounds of democratic theory could appeal to the standards criterion in demanding more stringent legislative directives.

In a practical sense, the standards criterion is difficult to apply and provides little restraint on legislative delegations. As the scope of governmental activity expanded, administration became more specialized and complex and less amenable to uniform standards set by the legislature. Delegations of authority either left a certain range of choice open to the administrator in implementing statutes or provided such discretion by default—through ambiguity or vagueness. Thus, considerable discretion accompanied delegations in new areas of governmental intervention. Further, it is doubtful that the balance of member incentives within the legislature favors greater specificity in lawmaking. The majority-building process demands frequent compromise and conciliation, often symbolized in vague statutory language that encourages consensus. Each demand for specific provisions and standards represents a political cost that must be weighed against chance for a bill's passage. Controversial demands also incur a potential cost in the political support of members' clientele. Through broad delegations of power, the costs inherent in resolving conflict among values and contending interests are passed on to the administrative agency charged with implementation. Legislators, then, have the opportunity to initiate reform, take credit, or assign blame, depending upon the eventual outcome.

THE PROBLEM OF DISCRETION

Within this climate of permissive delegation, there was some concern for fairness in administrative dealings with the public. The delegation of legislative and adjudicative powers to administrative agencies, while weakening the responsibility of the legislature to the public, had created a new relationship between the agencies and specific clientele groups. The substantive accountability of legislators for policy that imposed duties and obligations on the public was eclipsed by a functional accountability that placed the agencies between the legislature and the public. The reaction against these developments focused, not on the character of legislative delegations, but rather upon agency performance. The Administrative Procedures Act of 1946 formalized agency proceedings within a legal framework, providing the courts with a number of procedural and substantive requirements, such as hearings and reasoned consistency in decision making, in order to control the exercise of agency discretion (Stewart, 1975). Nonetheless, the overall impact of this pattern of litigation upon administrative accountability was limited to the satisfaction of private plaintiffs' claims against agency sanctions. In other words, the imposition of due process safeguards formalized adjudication, but was of no consequence for the agency's legislative prerogative.

A potent rationale for broad delegations of legislative power to sustain administrative discretion is found in pluralism (Herring, 1936; Dahl, 1961). The exercise of discretion involves balancing the competitive claims of organized interests, who are affected by policy. The administrator's selection and weighting of value premises emerges from the process of bargaining and accommodation among these groups. Broad delegations of authority place few restrictions on this process by leaving implementation open to group resolution. Bargaining among groups displaces formal legislative processes, with policy-making power divided among contending interests. The agency takes the place of the legislature as the focal point of group pressures. The legitimacy of the outcomes is based both on the perceived legitimacy of the process and upon the notion of representation and equal access. Vague legislation provides those who lost in the formulation phase with a second opportunity for influencing policy.

Without standards and directives to guide implementation, the administrative agency is subject to a variety of pressures from its clientele seeking favorable decisions. As Lowi (1969) points out, bargaining over implementation is simply logrolling on a case-by-case basis to determine the impact of decisions on interested groups. One result is the separation of the program to be administered from elective political responsibility; the product of interest logrolling is functional accountability and cooptation. The dependence of the agency on these groups is accentuated, and a systematic bias enters agency policy. Not surprisingly, agency policy that emerges from bargaining is most responsive to powerful clientele. Agency rescurces, including political clout, funds, and personnel, are limited relative to those of organized groups. Rather than a costly adversary posture, agencies draw useful information and support from these external sources.

Cooptation occurs where structural devices, such as decentralization, are used to facilitate access to bargaining; the results are inherently conservative. A good example of this is the Community Action Program.[2] Implementation was to be accomplished through the involvement of local organizations and of groups created with government assistance. Nonetheless, power was not shared along with access. Again, the outcomes of bargaining reflected the values of the dominant vested interests. The previously unorganized were brought into the process and were given an official status which separated them from those remaining outside. The situation not only stifled the emergence of newer groups (Lowi, 1969:245), but also restricted the scope of their demands (Peterson, 1970:501) and created a buffer between citizens and officials which legitimated the existing structure of power (Katznelson, 1972:333). The accountability characterizing this process is neither functional nor democratic, but responds to changes in the balance of power among interests. Thus, access without formal guarantees to compensate for resource disparities merely reinforces a conservative bias in policy.

In responding to the legislature's delegation of responsibility for value choices, the administrative agency is drawn into a surrogate political process. The representational linkage, placing elected officials between the public and administrators, is subordinated to the agency's direct interaction with organized groups. The legitimacy of selection supposedly is derived from the process of accommodating disparate interests. However, the process remains informal and subject to shortcomings that undermine any legitimacy in its outcome. The following sections consider institutional responses to this problem, focusing upon the control of administrative discretion.

JUDICIAL CONTROLS AND REMEDIES

The right of the individual to appeal to the courts against administrative action can be a potent check on agency abuses. Judicial review focuses on the legal authority of agency decisions and upon the fairness of associated proceedings. However, the scope and availability of review, determining its practical significance, have only recently been expanded. Under the traditional pattern of control, the courts were seldom consequential for enforcing agency responsiveness. Many of these limitations have been overcome in developments aimed at remedying the bias in the exercise of agency discretion.

THE TRADITIONAL LIMITATIONS OF COURT ACTION

Only a small percentage of instances where an individual seeks redress for adverse administrative action eventually comes under the scrutiny of the courts. Most are settled through informal agency accommodation. A few cases are the subject of formal administrative proceedings, a fewer still are appealed to the courts. The chief function of review was not to evaluate the content of agency decisions, but only to curb abuses in agency power. Judicial controls provided an inadequate remedy for the biases in administrative discretion. Review was available in only a limited number of cases, following the completion of administrative proceedings. The cost and delay involved in court litigation were effective obstacles to many who might otherwise seek judicial remedies. In any event, courts defer to the agency's decision, if it has some reasonable basis, despite possible disagreement over the conclusions. The court's determination is restricted to whether the agency acted within its power and whether its judgment was supported by substantial evidence (Gellhorn, 1972:263-269).

In the past, due process protection was restricted to parties whose property or liberty rights were subject to unauthorized governmental interference, usually in the form of imposed sanctions or obligations. These parties were entitled to initiate formal agency proceeding or to seek judicial review (Gellhorn and Byse, 1960:218-225). Under this scheme, opponents of the sanctioned party's position, and other individuals or groups with an interest in the outcome, were excluded from all but the informal proceedings. Agency dependence on clientele groups for information in proceedings and reluctance to undergo expensive litigation were conducive to favorable decisions. The value premises were set by economically powerful groups, as agencies remained unresponsive to the less organized.

The concern here is not exclusively with cases in which powerful clientele groups fight the imposition of agency sanctions through litigation, but also with the more frequent instances of agency bias in statutory enforcement. Interests of those, not subject to sanctions but rather intended as beneficiaries, could be disregarded through selective underenforcement, with claims against agency decisions dismissed in informal proceedings. Recent court action has expanded due process safeguards from the traditional protections of private rights against agency encroachments to the protection of advantageous relations with the government, including welfare entitlements, tenancy in public housing, scholarships, and government employment (Stewart, 1975:1717-1722). This development is consistent with the changes in standing requirements to include the intended beneficiaries of agency enforcement and regulation, particularly those who are adversely affected by agency discretion.

A REMEDY IN INTEREST REPRESENTATION

In effect, the court has changed its orientation from protecting the property and liberty of a small class of interests to one of recognizing substantive interests apart from any administrative illegality (Orren, 1976:731). The court addressed the agency bias, in favor of organized interests, and unresponsiveness to the unorganized by assuring broad representation of all affected interests in the agency's exercise of delegated powers. This included an expanded view of rights to initiate and participate in agency proceedings and the elimination of traditional barriers of standing to seek judicial review.

Previously, standing to intervene in agency proceedings was limited by the same principle that applied to standing to seek judicial review; parties had to demonstrate a substantial personal or economic stake in the outcome. The first sign of judicial support for expanded participation came in *Scenic Hudson Preservation Conference* v. *FPC* (1965). The court overturned the FPC's rejection of evidence offered by conservation groups, who opposed the siting of a power plant, and held that requirements of standing for intervention and review were satisfied by aesthetic, conservational, and recreational interests. It also imposed an obligation on the agency to consider the issues raised by the intervenors in a form that would be adequate for subsequent review. The Supreme Court, in two 1970 cases, expanded the scope of standing further to include those whose interests are "Arguably within the zone of interests to be protected."[3] Since this principle was based on a provision of the Administrative Procedures Act, it was no longer necessary to demonstrate that interests affected by an agency decision were legally protected by a relevant statute. Of course, the terms "arguably" and "zone" are open to interpretation and thus to judicial discretion (Gellhorn, 1972:254).

Implicit in the judicial expansion of participation is the recognition that the pluralist process can provide a legitimate framework for agency decisions and an acceptable solution to the problem of discretion, as long as appropriate consideration is given to unorganized, public interests (Stewart, 1975:1715). The accomplishment of this task has drawn the court into weighting the substantive claims of various interests to ensure adequate protection and full agency consideration. From the agency's perspective, the direction of judicial revisions is difficult to predict; the scheme of interest representation is built on a foundation of judicial discretion over access. Furthermore, there is no guarantee that the court's procedural adjustments will improve the eventual outcome. Despite an increase in the number of sources providing information, the agency has little incentive, other than fear of occasional judicial remand, for changing its pattern of decision making. Compensating for biases in a pluralist process is not only exceedingly expensive, but also ignores the fundamental issue of the delegation of legislative power. At best, a procedural solution to the problem of discretion will result in accommodation of affected interests–a restricted range of policy values. Nonetheless, procedural safeguards may provide the only avenue of representation for "public interest" claims.

Intervention and standing for review are rights separate from any pertaining to the initiation of agency action. If the agency does not institute enforcement proceedings or dismisses controversies in informal proceeding, standing to intervene or seek review are of no consequence. As Davis (1969) notes, agency "no action" decisions and informal settlements traditionally have not been subject to review, unless they were followed by formal proceedings. Clearly, this represents a serious flaw in the judicial scheme of protection for the rights of intended beneficiaries. If the agency is able to exclude selected groups from informal proceedings, this disarms any impact these groups might have at later stages. The crucial accommodations could be made informally and merely validated in formal proceedings. Recently, the courts have taken some initiative in controlling instances of discretionary nonaction by scrutinizing the agency's informal choices.[4] This further expansion of review has afforded the right of those with a stake in the agency's decision to initiate and participate in informal proceedings with settlements subject to appeal. Not only is this a necessary supplement to other safeguards, but it is vital to any notion of agency accountability in the face of congressional laxity. If one ignores the theoretically preferable solution to the problem of agency discretion based upon legislative reassertion, judicial controls provide a practicable mechanism of accountability.

LEGISLATIVE OVERSIGHT

THE STANDARDS DOCTRINE

In principle, the legislature is responsible for making choices among contending values and embodying these choices in statute. The legitimacy of this value-setting function is based upon the electoral consent of the public. Within this scheme, any administrative action affecting the public must be authorized by the legislature. Moreover, any public monies spent for these activities must first be appropriated through legislative enactment. The agencies are dependent upon the legislatures for their enabling legislation, their staff, and function, as well as for the scope and level of their program activities. Delegations of power to the agencies would employ the cooperative effort of Congress and the Judiciary, Congress defining clear limitations and standards for agency activities, and the courts confining the agency to its delegated authority. This is taken by some (Lowi, 1969; Friendly, 1962) to be a model for securing administrative accountability.

Recognizing the institutional constraints on congressional ability to specify meaningful standards, Lowi (1969:297-299) calls for the court enforcement of the "standards" doctrine of delegation, whereby delegations of authority to agencies, not accompanied by clear standards of implementation, would be invalidated. The legislature would be compelled by potential judicial sanction to restrain agency discretion. This assumes that both the judiciary and the legislature are able to determine the requisite specifications of clear standards of implementation, apart from any willingness to apply them. A less optimistic position is advanced by Davis (1969:27-51), who would lodge the task of specifying standards within the agencies, rather than within the legislature. The agencies would be charged with the formulation of rules to limit their exercise of discretion. Although this view absolves the legislature from structuring agency discretion, emphasis is still placed upon the role of the courts in determining the adequacy of the agency's self-imposed restraints.

Contemporary developments in judicial control have not moved toward the enforcement of either agency or legislative standard-setting practices, but rather toward interest representation in discretionary decision making by the agencies. Likewise, legislative attention to the control of agency discretion has focused upon piecemeal supervision of agency decision making and assessment of performance, rather than upon the systematic articulation of standards to guide implementation. On one hand, the resolution of values by the legislature, along with the provision of standards, is itself an important value in our political system. On the other hand, it is unrealistic to assume that values set in the legislature will be any less oligarchic than values set within the agencies, especially in light of judicial remedy of the most serious agency abuses. In calling for strict imposition of standards, instead of channeling and supervision of discretion, it is instructive to inquire to whose values legislative controls make administrators responsive. Clearly, the administrator also should be responsible to parties other than committee chairmen, such as to those directly affected by his decisions–public interest groups, intended beneficiaries, and clientele. A return to the standards doctrine of delegation offers a solution, albeit an intractable one, to the pluralistic bias in agency choices, but it does so at the cost of administrative judgment. Since there are no guarantees that exchanging detailed legislative directives for administrative judgment will be substantially better than policy choices, it is difficult to justify extinguishing the political process associated with agency discretion, so that the legislative process might predominate. Just as the representational flaws within the legislative system can be ameliorated through a supplementary political process with judicial sanction, so can the procedural and substantive flaws in this supplementary system be remedied through effective legislative oversight.

MOTIVATION AND RESOURCES

There is a continuum of oversight and involvement in agency activities that Congress is capable of exercising; it ranges from informal inquiries to statutory directives, from casual supervision to systematic control. As distinct from the judicial focus on "cases and controversies," the legislature has responded to the problem of administrative discretion by relying principally on various mechanisms to monitor activities and to assess program performance. The transition from monitoring to manipulation usually follows a loss in confidence either through abuses of committee trust or dramatic evidence of failure. Other levels of oversight activity correspond to varying levels of legislators' motivation and interest. Although few members would deny the importance of oversight, as an investment of legislators' time, it shifts in priority relative to other activities (Ogul, 1973). The incentives for oversight are strongest when such activities would complement concern for political survival. Scher (1963) has identified a number of strategic conditions that stimulate oversight, premised upon a cost-benefit calculus of political consequences. In this view, a commitment to oversight is contingent upon the opportunities for political gain. Surely, with the range and frequency of oversight activities, this is too Machiavellian a requirement. Responsibility to constituents and beneficiaries, as well as program interest and commitment, are likely to affect the priority and extent of a legislator's oversight activities.

Although legislators' motivation for oversight varies with levels of programmatic concern and confidence in administrators, the institutional commitment of resources for such activities has increased substantially in the last few years. The legislative mandate for oversight of administrative agencies was reviewed in the 1970 Legislative Reorganization Act: quoting from the relevant section, each standing committee shall "review and study, on a continuous basis, the application, administration, and execution of those laws, or parts of laws, the subject matter of which is within the jurisdiction of that committee." This commitment was expressed further in a requirement that each committee report on its review activities at the end of each Congress. In addition to enlarging committee staffs, the Act tied the General Accounting Office and the Congressional Research Service to oversight activities by authorizing them to review and analyze program activities (Ripley and Franklin, 1976:176). The Committee Reform Amendments in 1974 strengthened the reporting requirements of committees and placed the Government Operations Committee in the House in charge of the oversight activities of other House committees. In that same year, a House Resolution made adjustments in the committee system, encouraging committees to set up special subcommittees to engage in oversight investigations (Jewell and Patterson, 1977:458). Finally, the Congressional Budget and Impoundment Control Act of 1974 substantially increased Congress' capability for performing program reviews and evaluation by creating the Congressional Budget Office and two generously staffed budget committees.

Two consequences of these developments can be identified. First, the growth in specialized staff is making it possible to sustain a continuing effort at program oversight without diverting other resources. The legislator is no longer compelled to choose between formulation and review activities. Although each may assign a different relative priority to review, a certain level of review will be maintained on a continuing basis; it is no longer reserved as a response to crisis situations. To some extent, the review infrastructure will create a need for more program oversight. The incidence of evaluation and review requirements contained in enabling legislation will increase. The Water Pollution Control Act Amendments and Water Resources Development Act are exemplary; they also suggest the second consequence of staff commitment to continuous program oversight—the standards of review have changed.

In the past when legislators undertook program oversight in response to a loss in confidence, a crisis, or an opportunity, the problem was one of political accountability; the program or its administrators were not meeting their value expectations. The values of the specialist performing an evaluation, however, are instrumental rather than political. Goals are imputed, and objective rules and criteria are applied to program characteristics. The assessment, especially if mandated in enabling legislation, will be given considerable weight relative to political values. Ironically, the legislature's inability or unwillingness to specify goals and standards provided the rationale for initial delegation to the agencies. Now, the legislature is evolving a bureaucracy to meet the agencies on their own terms. In other words, expertise is seen as both the ailment—the *raison d'être* of agency discretion—and the cure. Rather than channeling discretion by defining and redefining the political component in agency decisions, evaluation and program review provides an institutional defense against the failures in agency performance. Accountability, then, is based on responsiveness to specialized performance values.

PROGRAM OVERSIGHT

The most extensive mechanism of supervision and control over agency activities is the legislature's authorization and appropriation of funds. The Appropriations Committees annually scrutinize the details of agency expenditures; the agencies, in turn, must account for their past use of funds and justify future uses. Although fiscal oversight has become more systematic since the adoption of procedural reforms in 1974, the process remains highly decentralized with autonomous subcommittees investigating selected agency practices in detail. The hearings before these subcommittees are conducive to accountability; the agencies exchange information and rationale for a clarification of legislators' views.

The policy views expressed during these hearings, or in the language of the committee's subsequent report, are treated as binding by the agencies, since the subcommittee can exercise a budget sanction for noncompliance. However, because the process is fragmented, agencies are likely to receive conflicting interpretations of legislative directives. The tendency for agencies fearing reprisals is to rely upon the interpretation of influential appropriations subcommittee members, even though these views may not be a legitimate representation of sentiment in the chamber. Furthermore, the conflict between authorizing and appropriations committees over agency policy creates distortions in accountability. The authorizing committees often revert to techniques that allow agencies to evade appropriations scrutiny. Entitlement provisions and contract and borrowing authority create spending obligations without the concurrence of appropriations committees. Conversely, earmarking funds for select purposes and placing substantive conditions on spending authority circumvent the authorizing committees. Thus, oversight takes on an oligarchic character, and accountability to one group is undermined by pressures for accountability to its competitor. The result is responsiveness, not to the legislature, but to a few individuals in strategic positions. Yet, there is also a concommitant reduction in agency discretion. The bias entering agency policy is that of subcommittee chairmen. If one has confidence in the responsibility of these individuals, then expansion of the scope of fiscal remedies and review is a desirable solution to the problem of agency discretion.

An additional mechanism of program oversight is the sunset statute.[5] This legislation provides for the automatic termination of program and budget authorizations at the end of a designated Congress, making reauthorization mandatory for the continuation of a program. Programs would be grouped by function and subject to scrutiny and possible reauthorization every sixth year. The objective is to force a comprehensive review of program commitments, while allowing for comparisons among similar authorizations. Periodic review and potential termination return the control of multi-year and permanent authorizations to the committees. Thus, the opportunity for regular oversight is distributed more evenly between authorizing and appropriations subcommittees. Although such a procedure undoubtedly would involve the agencies and their clientele in desperate rationalizations and symbolic reassurances, the idea of a comprehensive review is appealing. Of course, its success depends upon the motivations and objectives of those implementing it. Nonetheless, a conscientious reassessment of commitments could strengthen accountability linkages between Congress and the public, as well as between Congress and administrative agencies. Moreover, the values implicit in assessments would not be strictly performance oriented, since review would be comprehensive and comparative, evaluating information beyond simple output measures.

REVIEW OF AGENCY DECISIONS

Another focus of oversight, more akin to judicial concerns, is upon administrative decision making. Two procedures are being used more frequently; one is a mechanism for prior review, and the other for post review. Prior review is implemented in requirements for legislative or committee clearance of a proposed decision, before it is executed. This is the only measure, apart from consultation, that permits the administrator to validate his delegated authority and the scope of his discretion. The approval of a decision may require either assent or the absence of disapproval. The latter is often referred to as the legislative veto. The second procedure, a post-decision review, consists of reporting requirements. There are two kinds of requirements: evaluations and recommendations, and descriptions of activities. Requirements often contain provisions for surveys and studies with quarterly or annual submission. In other cases, continuous reporting is mandated. For example, the Energy Reorganization Act of 1974 states, "The Administrator shall keep the appropriate congressional committees fully and currently informed with respect to all of the Administrator's activities." This represents a commitment to continuous surveillance of agency activities.

Once restricted to reorganization and to public works projects, the legislative veto has recently been applied to agency decisions regarding implementation standards (Bruff and Gellhorn, 1977). This device counterbalances broad delegations with supervision of policy initiatives. Veto provisions may be included in legislation either because of past abuses and mistrust, or due to the controversiality of the issues at stake. Although the objective is to enforce the intent of the legislature in agency rule making, the effect is to defer questions of policy to negotiations between agency and committee staffs. Thus, the committee is likely to shape intent to its own purposes and to the need for conciliation. Moreover, such complicity would preclude the possibility of judicial remedies. For judicial and legislative controls to be most effective, their complementaries must be nurtured rather than ignored.

SUPPLEMENTING EXTERNAL CONTROLS

A great deal of the legislator's time and staff resources are devoted to casework for his constituents. Most cases involve requests for administrative intercession in their behalf, either to obtain materials and information, employment, or redress (Gellhorn, 1966). The typical response to these requests is a letter or telephone call of routine inquiry. Although a legislator's referral may draw a quick response, satisfaction–especially in cases involving maladministration–is the exception. Part of the problem in treating serious grievances is that the legislator's casework function operates separately from the more potent oversight activities of the committees. The relevance of individual complaints to the larger issue of accountability is lost, since there is no attempt to establish systematic patterns of abuse. The ombudsman idea is useful in this context: an Administrative Counsel could be appointed to establish a central clearinghouse for individual cases to avoid duplication of effort (Gellhorn, 1966:89) and to bring the legislature's oversight mechanisms to bear in areas of administrative abuse. Each legislator would maintain his service relationship with constituents, while diffused pressures for reform would be aggregated as a need for accountability. Rather than supplant the casework function, a central clearinghouse would integrate incidences of citizen abuse by administrative agencies into the over-all framework of legislative oversight.

NOTES

1. Both struck down sections of the NIRA, *Panama Refining Co.* v. *Ryan* 293 U.S. 388 (55 S. Ct. 241) and *Schechter Poultry Corp.* v. *U.S.* 295 U.S. 495 (55 S. Ct. 837), five months apart in 1935.

2. The Community Action Program, established by the Economic Opportunity Act of 1964, mandated "maximum feasible participation" to reduce discrimination in implementation.

3. These were *Association of Data Processing Service Organizations* v. *Camp,* 397 U.S. 150 (90 S. Ct. 827) and *Barlow* v. *Collins,* 397 U.S. 159 (90 S. Ct. 832), both in 1970. The Court reduced the law of standing to two questions: (1) Is the complainant "aggrieved in fact," and (2) Is the interest sought to be protected "arguably within the zone of interests to be protected by the statute in question?"

4. In *Environmental Defense Fund* v. *Ruckelshaus,* 439 F. 2d 584 (D.C. Cir. 1971), the court required the agency to rely on formal rather than informal proceedings to determine the safety of DDT. In so doing, they recognized that informal proceedings were unfairly favorable to powerful economic interests. See also *Medical Committee For Human Rights* v. *SEC,* 432 F. 2d 659 (D.C. Cir. 1970) for court treatment of administrative nonenforcement. Despite these inroads, cost remains a formidable barrier to court intercession. Transcript and attorney fees set the minimum level at about $3,000, depending on venue and complexity of one's claim. Furthermore, there is no assurance that the presiding judge will be sympathetic with these developments.

5. Successive versions of a sunset bill have been reported by Senate committees. The most recent version (S 2), introduced by Senator Muskie at the beginning of the 95th Congress, has bipartisan support in the Senate. With an automatic termination, Congress would be forced to consider the reauthorization of programs—either to continue them or to let them "fade into the sunset." The term "sunset review" replaced the less glamorous "zero-base review."

REFERENCES

BRUFF, H., and GELLHORN, E. (1977). "Congressional control of administrative regulation." Harvard Law Review 90(7): 1369-1440.

DAHL, R. (1961). Who governs. New Haven, Conn.: Yale University Press.

DAVIS, K. C. (1969). Discretionary justice. Urbana, Ill.: University of Illinois Press.

FRIENDLY, H. (1962). The federal administrative agencies. Cambridge, Mass.: Harvard University Press.

GELLHORN, E. (1972). Administrative law and process. St. Paul, Minn.: West Publishing.

GELLHORN, W. (1966). When Americans complain. Cambridge, Mass.: Harvard University Press.

---, and Byse, C. (1960). Administrative law, cases and comments. New York: Foundation Press.

GILBERT, C. (1959). "The framework of administrative responsibility." Journal of Politics, 21(2):373-407.

HERRING, P. (1936). Public administration and the public interest. New York: McGraw-Hill.

HYNEMAN, C. (1950). Bureaucracy in a democracy. New York: Harper.

JEWELL, M., and PATTERSON, S. (1977). The legislative process in the United States (3rd ed.). New York: Random House.

KATZNELSON, I. (1972). "Antagonistic ambiguity." Politics and Society, 3(1):323-333.

LOWI, T. (1969). The end of liberalism. New York: Norton.

OGUL, M. (1973). "Legislative oversight of bureaucracy." Paper prepared for the Select Committee on Committees, U.S. House of Representatives, 93rd Congress, 1st session, Washington, D.C.: Government Printing Office.

ORREN, K. (1976). "Standing to sue." American Political Science Review, 70(3):723-741.

PENNOCK, J. (1960). "The problem of responsibility." Pp. 3-27 in C. Friedrick (ed.), Responsibility (Nomos III). New York: Liberal Arts Press.

PETERSON, P. (1970). "Forms of representation." American Political Science Review, 64(2):491-508.

RIPLEY, R., and FRANKLIN, G. (1976). Congress, the bureaucracy, and public policy. Homewood, Ill.: Dorsey.

Scenic Hudson Preservation Conference v. Federal Power Commission (1965). 354 F. 2d 608.

SCHER, S. (1963). "Conditions for legislative control." Journal of Politics, 25(3):526-551.

SIMON, H., SMITHBURG, D., and THOMPSON, V. (1950). Public administration. New York: Knopf.

STEWART, R. (1975). "The reformation of American administrative law." Harvard Law Review, 88(8):1669-1813.

9

Legislative Accountability: Electoral Margins, District Homogeneity, and the Responsiveness of Legislators

JAMES H. KUKLINSKI

□ THE GROWTH of a diverse and heterogeneous society in the United States has posed difficult problems for democratic governance. As modern society evolved, totally participant democracies such as the New England town meeting quickly became vestiges of the past (Luttbeg, 1974:1-2). Wishing to retain the moral basis of popular government embedded in these direct democracies, theorists, and practitioners alike searched for and instituted mechanisms which would ensure that popular preferences control at least the basic direction of public policy. If officeholders could be made accountable to the mass public, framers concluded, basic democratic principles would not be lost.

The primary mechanism for ensuring accountability in a representative system of government is the election. "The democratic method," Joseph Schumpeter (1947:269; see also Dahl, 1965) states in his classic, *Capitalism, Socialism and Democracy,* "is that institutional arrangement for arriving at political decision in which individuals acquire the power to decide by means of a competitive struggle for the people's vote." Because individuals desire to

AUTHOR'S NOTE: *The insightful work of Morris P. Fiorina has influenced substantially my thinking about the representational linkage. I would like to acknowledge my intellectual debt here. He is not, of course, responsible for any errors in my own work. Written communication from Heinz Eulau regarding earlier work also proved useful in developing the ideas expressed in this chapter.*

gain and retain public office, they must pay due attention to the desires of the electorate. In this fashion accountability of the few to the many is guaranteed.[1]

By extension the greater the competition, i.e., the greater the probability of electoral accountability, the more likely a representative will feel compelled to respond to those whom he represents.[2]

> legislators from marginal or competitive districts will be more responsive to their constituents than those elected with minimal challenge. . . . [L] egislators want to get re-elected and that competition poses a threat to continued tenure. Since he must seek electoral support from voters in his district, the representative acts continuously to maximize that support. He is especially conscious of his roll-call voting record because he is convinced of the visibility of that record to his constituents. [Jones, 1973:925-926]

The relationship between the policy preferences of the electorate and the policy acts of the representative, in other words, should increase as the winning margin of the representative diminishes.

Because it is at the core of representational theory, a number of political scientists have attempted to ascertain empirically whether representatives who are elected by small electoral margins–and thus are potentially more electorally accountable–in fact do respond more readily to their constituencies than representatives who win by large margins. Even though this question is seemingly congenial to systematic data analysis, however, we simply cannot say with confidence that the "marginality hypothesis" is or is not correct. Findings have been confusing and contradictory; at best we can say that the hypothesis is not as easily testable as once was believed.

Part of the confusion emanates from the fact that researchers have overlooked the question of responsive *to whom?* In districts which are very homogeneous with regard to policy preferences, the relevant "whom" is probably the large majority which agree on the policy question. But as a district becomes more heterogeneous, the "to whom" question increases in complexity. Significant differences of opinion on policy questions exist; and consequently the representative from a heterogeneous district finds himself facing a more difficult situation than a representative from a homogeneous district. Responsiveness in this instance may not be a simple matter of doing what the majority desires. The representative may see himself as accountable to diverse and multiple clienteles.

The purpose of this chapter falls far short of providing "conclusive" evidence about the conditioning influence of district marginality on the responsiveness of representatives. As a more modest task, we raise questions about the line of reasoning which currently seems to be accepted by students of legislative behavior and elections. Our purpose is not to reject this stream of argument, but rather to make it more precise. Although we present data, we do so only in terms of exploration. Our spirit is one of advancing scholarship on a topic basic to the theory and practice of democratic government.

THE MARGINALITY HYPOTHESIS AND DISTRICT HOMOGENEITY

Early investigations of the marginality hypothesis[3] appeared to lend support to it (MacRae, 1952; Dye, 1961; Patterson, 1961; Pesonen, 1963; Shannon, 1968; LeBlanc, 1969). Employing party loyalty indices as measures of representative behavior, these studies generally found that marginal district representatives were more likely than those from safe districts to abandon their parties and cast votes with the opposition party. Thus, researchers concluded, legislators in situations of high electoral accountability pay greater attention to the policy preferences of their constituencies than do legislators in situations of low electoral accountability.[4]

Unfortunately these studies can be criticized on the grounds that there is no valid reason for assuming that increased party disloyalty to a representative's party implies increased loyalty to his constituency (Fiorina, 1973; 1974). A representative may bolt his party for a number of reasons, of which constituency is only one. In fact "deviation from party does not even imply that the representative deviated *in the direction of* constituency pressure" (Jones, 1973:934). The extent to which a legislator remains loyal, furthermore, may be dependent upon the extent to which party majority position and district attitude coincide. As we found in earlier research (Kuklinski, 1977), increased disloyalty among California assemblymen led to greater policy agreement on a taxation dimension, but not on a contemporary liberalism dimension. On the latter policy domain, majority position for both parties was determined by their respective constituencies; thus those who remained loyal to their parties also served the wishes of their constituencies. In this instance, increased party disloyalty clearly did not imply a stronger relationship between a representative's roll-call behavior and his district's policy preferences.

Given these legitimate criticisms, we are left with rather meager evidence regarding the possible conditioning influence of district marginality on representative behavior. Undoubtedly the most persuasive findings are those derived from the 1958 SRC Representation Study. Contrary to the earlier findings, Miller (1970) provides evidence that the representational linkage is stronger in safe than competitive districts. Using a direct measure of con-

TABLE 1
HOUSE WINNERS' PROPORTIONS AS A FUNCTION OF DISTRICT ATTITUDINAL HOMOGENEITY

Variable[a]	B	δB	R^2	a	s.e.e.
CR var	−.14	.06	.16	.71	.17
SW $\overline{m}$	.30	.15			

[a]CR var is the variance of each constituency sample around that district's mean attitude on the civil rights domain. SW $\overline{m}$ represents the proportion of each district sample which occupies the mode of the attitude distribution along the social welfare dimension. The remaining variance and modal measures are insignificant.

SOURCE: Morris P. Fiorina (1974). Representatives, roll calls, and constituencies. Lexington: D. C. Health: 98.

stituency attitudes on three policy domains—civil rights, social welfare, and foreign affairs—he shows that congressmen from safe districts not only are more representative than marginal district members, but they are also more likely to vote in accordance with their relatively accurate perceptions of district opinion. Marginal district congressmen, in contrast, are much less accurate in their perceptions and display a clear tendency to follow their own predispositions when voting.

Due in part to the availability of a constituency opinion measure derived from survey data,[5] Miller's findings have influenced subsequent reasoning about the impact of marginality on the opinion-policy linkage. This reasoning, which now is widely accepted, can be stated thus: It is much easier for legislators in relatively homogeneous districts than for those in heterogeneous districts to find consensus among constituents and thus to accurately identify district opinion. Consequently it is easier for them to be in agreement with a majority of their constituents. And, of course, it is in such districts that legislators are likely to be elected by safe margins.

This argument as presented portends a very minimal role for electoral competitiveness as an accountability mechanism. It purports that the distribution of opinion within a district, not its marginal character, specifies the connection between a representative's roll-call votes and district opinion. Intense political competition itself will not guarantee relatively high levels of policy agreement. Before confidently accepting a conclusion of such significance, however, we must be certain that the effect of district marginality on the representative behavior of legislators has been properly identified. More to the point, we need to ask: Are district marginality and homogeneity so closely related such that we can justifiably assume that one is simply an underlying dimension of the other?

We believe not. Consider, for example, Fiorina's (1974:89-100) analysis of the relationship between the two variables. Using two measures of homoge-

TABLE 2
A PROPOSED SCHEME TO MEASURE THE INDEPENDENT EFFECT OF DISTRICT MARGINALITY ON REPRESENTATIVE BEHAVIOR

Homogeneous Districts[a]		Heterogeneous Districts	
Competitive	Safe	Competitive	Safe
A	B	C	D

[a]Entries represent the level of policy agreement.

neity/heterogeneity for each of the three SRC policy domains, he finds the relationships shown in Table 1. Although several of the regression coefficients of the homogeneity/heterogeneity measures are significantly related to the marginal character of the district, the important point here is that not all competitive districts are heterogeneous, and not all safe districts are homogeneous. The magnitude of the standard error of estimate (.17) is particularly revealing; the average difference between the actual competitiveness of a district and that predicted by the homogeneity/heterogeneity measures is some 17 percentage points.[6]

Before any final conclusions can be drawn about the impact of district marginality, we may need to consider the scheme depicted in Table 2. If competitiveness has an effect on the behavior of representatives which is *independent of* district homogeneity, then there ought to be a difference between A and B on the one hand, and between C and D on the other. In terms of the marginality hypothesis, we expect that the level of representation in A-type districts will be greater than that in B-type districts because of the greater effort by marginal district representatives to pursue their constituencies' policy preferences. Similarly, the hypothesis predicts that policy agreement will be greater in C-type districts than in D-type districts.[7] By, in essence, removing any confounding effect of district homogeneity, we can begin to answer the criticism that "empirical research has failed to show . . . that the safety/marginality of districts has . . . an effect on voting above and beyond what might be caused by the homogeneity . . . of districts" (Jewell and Patterson, 1977:407).

DATA

A key element in examining the scheme we have just outlined is the availability of a direct constituency opinion measure. Without it we cannot be certain that we have adequately identified the extent of agreement between the policy preferences of the electorate and the policy acts of its representative.

In an effort to study other conditioning influences on the representational linkage, we explored elsewhere (Kuklinski, 1978) the possibility that voting returns in referenda and initiative issue could be used to measure district-wide opinion.[8] The growing consensus among scholars of representation that the connection between representative and constituency occurs through broad policy domains rather than specific issues (Miller and Stokes, 1963; Clausen, 1973; Stone, 1977) suggested the desirability of determining whether these issues could be arrayed along broader policy domains. Utilizing aggregate voting returns on propositions placed before the California electorate at the 1968, 1970, and 1972 general elections, we were able to determine, using the technique of factor analysis, that the issues on which the electorate expressed their opinions could be placed into three distinct domains: contemporary liberalism, taxation, and government administration. The standardized scores produced by the factor analysis allowed us to rank the state's assembly districts in terms of support for each domain.

Although scholars had shown that legislators' roll-call behavior could be understood in terms of dimensions sufficiently broad to comprehend a number of single votes, a compelling concern was whether dimensions equivalent to those identified among the constituencies could be found at the

TABLE 3
ISSUES COMPRISING THE CONTEMPORARY LIBERALISM DIMENSION

Issue	Factor Loading
Removal of state penalties for personal use of marijuana	.956
Redefinition of obscenity; elimination of "redeeming social importance" test	–.865
Authorization of motor vehicle fuel tax revenues for air pollution control	.853
Creation of Coastal Zone Conservation Commission to control coastal development	.821
Reimposition of the death penalty	–.800
Institution of prohibitions on labor relation activities, particularly as they apply to agricultural workers	–.788
Extension of the right to vote to 18-year olds	.725
Addition of section to Education Code eliminating mandatory busing to achieve racial balance	–.689
Reduction of requirement to pass bond issue for replacing unsafe public school buildings from two-thirds vote to simple majority	.657
Provision of a $250,000,000 bond issue to provide funds for water pollution control	.610
Issuance of revenue bonds to finance installation of pollution control facilities	.585
Elimination of property tax for welfare purposes	–.571
Addition of right of privacy to inalienable rights of people	.485

legislative level. Fortunately for our purposes, we were able to construct equivalent legislative dimensions for each of four years, 1970-1973 (see Kuklinski, 1978 for a fuller discussion), so that we could with reasonable confidence examine the connection between public opinions and public actions.

We shall focus on only one of the dimensions—contemporary liberalism. The relationship between the policy preferences of constituencies and the policy acts of representatives consistently has been stronger on this domain than on taxation or government administration. This can be attributed in large part to the relatively high saliency of contemporary liberalism issues to the electorate. Questions dealing with minority rights and changes in life-style have been at the forefront of California politics during the last decade. The relevant point here is, to the extent that the increased probability of electoral sanctions leads to increased responsiveness, its effect should be most noticeable on the contemporary liberalism dimension.

Using the thirteen issues which define the domain (Table 3), we computed a homogeneity/heterogeneity measure thus:

$$\text{Index of Homogeneity/Heterogeneity} = \frac{\sum_{i=1}^{N} X_i - 50}{N}, \text{ where}$$

X_i = percent of district vote in support of contemporary liberalism on proposition i

N = number of propositions comprising the dimension

The working assumption underlying this index is that the most heterogeneous district is one in which the actual vote in support of each proposition is 50%. The extent to which a district becomes increasingly homogeneous on the dimension[9] depends upon (1) the relative proportion of propositions on which the district is supportive to those on which it is nonsupportive, and (2) the margin by which the votes on the propositions deviate from 50%. Although far from perfect, this measure appears to be similar to that which Fiorina computed from the SRC data. When combined with information in the electoral fortunes of incumbent assemblymen, we can explore the question of whether electoral competitiveness has an effect on the representational linkage which is independent of that due to the homogeneous or heterogeneous nature of a district.

FINDINGS

Are homogeneity/heterogeneity and safety/marginality sufficiently *un*related so that we can construct the combinations of district types delineated

Figure 1: THE RELATIONSHIP BETWEEN DISTRICT HOMOGENEITY AND ELECTORAL SECURITY, 1970 AND 1972

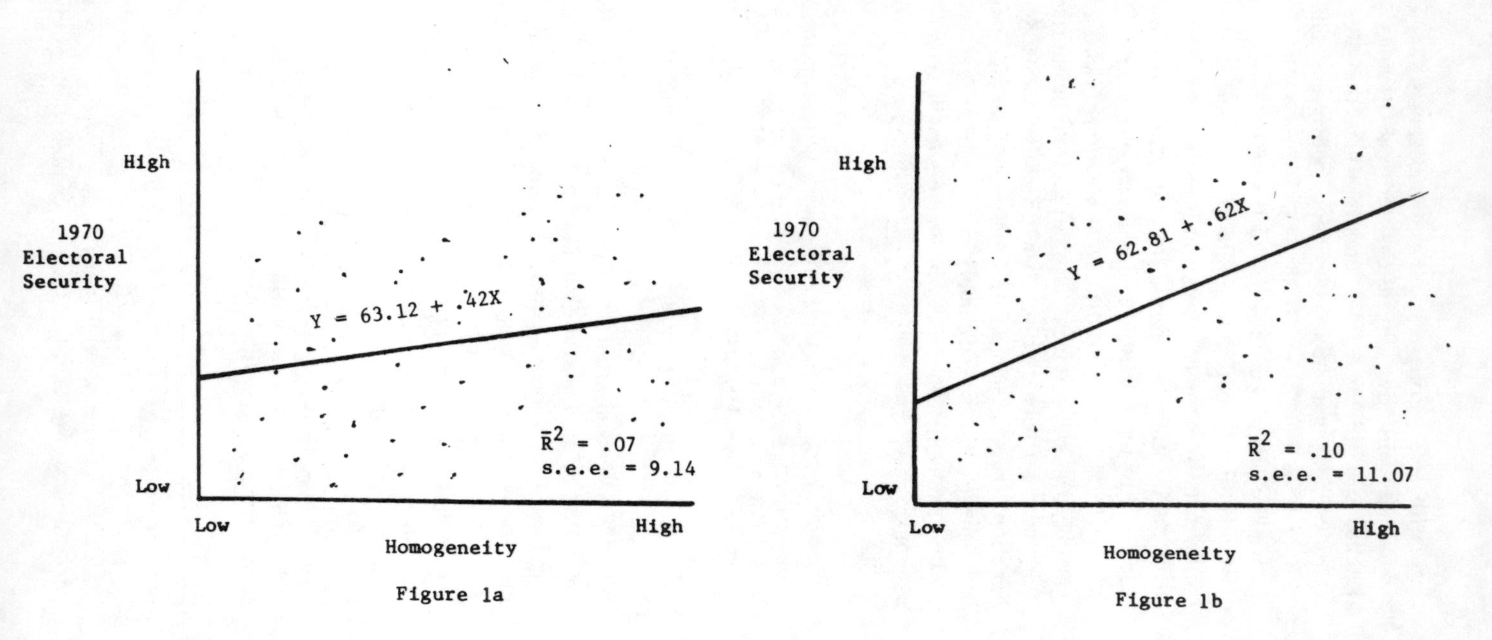

in Table 2? And if so, does the evidence support the hypothesis that representatives who find themselves in a situation of high electoral accountability act in a more responsive manner than representatives who face less threatening electoral sanctions?

We used bivariate regression analysis to specify the relationship between the winner's proportion of the vote in the 1970 and 1972 elections and our measure of homogeneity/heterogeneity. Figures 1a and 1b indicate that the variables are related in the expected direction: as the homogeneity of a district increases, the electoral security of its representative tends to rise. Two points deserve mention, however. First, homogeneity is a better predictor of the 1972 vote than the 1970 vote; the implication is that homogeneity/heterogeneity itself does not explain the marginal nature of legislative districts. Second, as was the case in Fiorina's analysis of the SRC data, the magnitudes of the standard errors of estimate suggest caution in assuming that safety/marginality is simply an artifact of homogeneity/heterogeneity. In one instance the standard error is 9.14; in the other it is 11.07. Homogeneity does not ineluctably imply safety; nor does heterogeneity without exception lead to marginality.

The addition of a dummy variable for incumbency (incumbent = 1, 0 otherwise) does not significantly change this conclusion. Using 1970 district competitiveness, the prediction with incumbency included is:

$$Y = 64.28 + .39X + 10.44I$$

$$R^2 = .22$$

$$\text{s.e.e.} = 8.34$$

For 1972 it is:

$$Y = 64.66 + 47X + 6.69I$$

$$R^2 = .15$$

$$\text{s.e.e.} = 10.78$$

Inclusion of the dummy variable improves the fit of the regression equations. The fact that homogeneity/heterogeneity coefficients for the two election

TABLE 4.
THE DISTRIBUTION OF LEGISLATIVE DISTRICTS ACCORDING TO COMBINATIONS OF HOMOGENEOUS/HETEROGENEOUS AND SAFE/COMPETITIVE

	1970			1972	
	Safe	Competitive		Safe	Competitive
Homogeneous	27	10	Homogeneous	30	8
Heterogeneous	26	12	Heterogeneous	24	13

years become very similar when incumbency is taken into account is particularly notable. The possibility of various district-type combinations nonetheless remains.

For the remainder of our analysis, therefore, we have classified assembly districts as shown in Table 4. Competitive districts are those in which the winner received 60% or less of the vote. This criterion has wide acceptance among legislative scholars. If a district's homogeneity/heterogeneity score falls above the median for all districts, we classified it as homogeneous; heterogeneous districts have scores below the median.

Although this classification of districts results in some loss of information, its utility lies in the fact that we now can look at the representational linkage within each combination. By comparing the representative behavior of legislators from competitive and homogeneous districts with that of representatives from safe and homogeneous districts, we can begin to ascertain whether the marginal character of a district has an independent effect upon the way representatives relate to their constituencies. We can similarly "control" the consensual nature of the district by looking at competitive and safe districts which also are heterogeneous.

The following dummy regression model achieves this purpose (Kmenta, 1971:409-430; Miller and Erickson, 1974):

$$Y = \beta_1 + \beta_2 X + \gamma_1(XD_1) + \gamma_2(XD_2) + \gamma_3(XD_3) + \alpha_1 D_1 + \alpha_2 D_2 + \alpha_3 D_3 + e$$

where

Y = roll call behavior [10]
X = constituency opinion
D_1 = 1 if homogeneous and safe (HoS)
0 otherwise
D_2 = 1 if heterogeneous and competitive (HeC)
0 otherwise
D_3 = 1 if heterogeneous and safe (HeS)
0 otherwise

This model partitions legislators into one of the four district types but uses all legislators in the estimation rather than the individual subsets.[11] The regression equation for representatives from each district type can be determined by looking at the appropriate regression coefficients. The equation for legislators from districts which are homogeneous and competitive (HoC districts) is given by

$$Y = \beta_1 + \beta_2 X$$

The equations for the other three subgroups then are

$$Y = (\beta_1 + \alpha_1) + (\beta_2 + \gamma_1)X \quad \text{(HoS)}$$

TABLE 5
RELATIONSHIP BETWEEN DISTRICT ATTITUDE
AND ROLL CALL BEHAVIOR BY DISTRICT TYPE

District Type	Intercept	Regression Coefficient	Significance of Difference Between Coefficients
1970			
Competitiveness			
HoC	−.08	.64	
HoS	−.11	.26	H_o: $\beta_2 = (\beta_2 + \gamma_1)$, $p < .05$
HeC	−.25	.17	
HeS	+.02	.32	H_o: $(\beta_2 + \gamma_2) = (\beta_2 + \gamma_3)$, $p < .05$
			$\bar{R}^2 = .31$
			s.e.e. = .28
1972			
Competitiveness			
HoC	.17	.39	
HoS	−.01	.16	H_o: $\beta_2 = (\beta_2 + \gamma_1)$, $p < .05$
HeC	.12	.05	
HeS	−.16	.27	H_o: $(\beta_2 + \gamma_2) = (\beta_2 + \gamma_3)$, $p < .05$
			$\bar{R}^2 = .34$
			s.e.e. = .22

$$Y = (\beta_1 + \alpha_2) + (\beta_2 + \gamma_2)X \quad \text{(HeC)}$$
$$Y = (\beta_1 + \alpha_3) + (\beta_2 + \gamma_3)X \quad \text{(HeS)}$$

We are particularly concerned with the relative magnitudes of β_2 and $(\beta_2 + \gamma_1)$ and, similarly, $(\beta_2 + \gamma_2)$ and $(\beta_2 + \gamma_3)$. The marginality hypothesis predicts that $\beta_2 > (\beta_2 + \gamma_1)$ and $(\beta_2 + \gamma_2) > (\beta_2 + \gamma_3)$.

We have completed a regression analysis for each of the two time periods. The results are presented in Table 5. The two equations are remarkably similar. In both instances policy agreement is higher in HoC districts than HoS districts, as the hypothesis predicts. The differences in the slopes of the two subgroups in fact are quite substantial. Contrary to original expectations, though, policy agreement is *not* greater in HeC districts than HeS districts. The relationship between constituency preferences and roll call votes is greater in HeS districts than HeC districts for both years. Among homogeneous districts increased marginality leads to a higher level of policy agreement; among heterogeneous districts representatives secure in their positions appear to be more in tune with the preferences of their constituencies than legislators from marginal districts.

We will postpone a substantive discussion of the findings to the next section. Suffice it to note at this time that electoral competitiveness appears to have an independent effect upon the behavior of legislators, and this influence is consistent across the two time points. We can emphasize the significance of this finding if we momentarily neglect the homogeneous/heterogeneous nature of a district and examine only the impact of district marginality on representative behavior, as researchers normally have done. The appropriate regression model then is

$$Y = \beta_1 + \beta_2 X + \gamma(XS) + \alpha_1 S + e$$

where

Y = roll call behavior

X = constituency opinion

S = 1 if safe

0 otherwise

The prediction for competitive district legislators then is given by

$$Y = \beta_1 + \beta_2 X$$

and that for safe district legislators by

$$Y = (\beta_1 + \alpha_1) + (\beta_2 + \gamma_1)X$$

The general equation for 1970 is

$$Y = -.05 + .16X + (-.05 - .18) + (.16 + .17)X$$

$R^2 = .24$

s.e.e. = .33

$H_o\colon \beta_2 = (\beta_2 + \gamma 1),\ p < .05$

For 1972 it is

$$Y = .09 + .11X + (.09 - .11) + (.11 + .16)X$$

$R^2 = .19$

s.e.e. = .35

$H_o\colon \beta_2 = (\beta_2 + \gamma_1),\ p < .05$

Using only electoral competitiveness, and thus not controlling for the nature of the district, we find results which are very similar to those reported by Miller. In both instances the representational linkage is stronger in safe than in competitive districts, i.e., $(\beta_1 + \gamma_1) > \beta_1$. By not controlling for the consensual nature of the district, in other words, we might incorrectly conclude that marginality is in no way related to the accountability of legislators.

DISCUSSION

The marginality hypothesis predicts that policy agreement will be higher in competitive than safe districts, presumably because the higher risk of electoral defeat will cause legislators from marginal districts to make a more sustained effort to be attuned to the political inclinations of their constituencies. We have found, examining two points in time, the prediction to be accurate only in districts which have a consensus of opinion. Among legislators from homogeneous districts, representatives who are electorally insecure are more likely to support policies in line with district majority preferences than are those who face less threat of electoral sanction. The explanation is seemingly simple enough. In homogeneous districts legislators can discern modal preferences with relative ease. Knowing unquestionably that it is the large majority to whom they are electorally accountable, representatives from HoC districts consistently respond to their policy preferences. HoC district legislators feel they cannot afford to ignore modal sentiment on most policy questions. Legislators who feel less threatened, in contrast, are sufficiently confident to eschew at times the operating principle of majority rule and respond to minority interests—whatever they may be (and which our data do not identify). As long as they don't ignore modal opinion on the most salient and visible issues, HoS legislators apparently have determined that they have some flexibility in their roll-call choices (see Fiorina, 1974:45).

The behavior of legislators from heterogeneous districts is more difficult to explain. Since policy agreement is higher in HeS than HeC districts, it is tempting to conclude that representatives from HeS districts are more responsive than representatives from HeC districts. As a logical proposition, however, it makes little sense to assume that representatives from HeC districts make *less* effort to please their constituents.

Our problem in trying to untangle this web is that we have a measure of policy agreement, but we do not have a measure of responsiveness (Eulau and Karps, 1977). The available data do not allow us to identify the efforts which legislators make to fulfill the policy preferences of their constituencies. Although any attempt at explanation can therefore only be speculative, we offer the following scenario as typical of behavior among legislators from heterogeneous districts.

Although they exist, majorities in heterogeneous districts are not particularly large. Differences in opinion on issues related to changes in life-style, moreover, are likely to be rather intense. Because of these configurations within the constituency, the legislator from a heterogeneous district faces what aptly has been labelled "mandate uncertainty" (Janda, 1961:169-179). Whereas determining the strategically correct response typically is a simple and straightforward matter for representatives from homogeneous districts, legislators representing heterogeneous constituencies face a less-than-obvious decision regarding the proper strategy to follow. They may side consistently with one group or, alternatively, they may try to balance the competing claims of the groups.[12] Our data suggest that HeC and HeS district legislators do not respond in the same way. Given the weak relationship between district-wide opinion and roll-call behavior, it would appear that the HeC district legislator sees himself as accountable primarily to one faction or group within the district. HeS representatives, on the other hand, apparently

TABLE 6
SENIORITY OF LEGISLATORS BY TYPE OF HETEROGENEOUS DISTRICT

Seniority	1971 HeC	1971 HeS	1973 HeC	1973 HeS
Median Years Served	3.8	7.1	2.5	8.0
Percent of Representatives Serving Two or Less Terms	33%	4%	61%	16%

are more inclined to balance the demands of the factions within their districts. Their roll-call behavior thus more closely reflects district-wide opinion.

It is important to emphasize again the tentative nature of our discussion. But if the scenario just described is correct, we ought to find that representatives from HeS districts are on the whole more senior than representatives from HeC districts. First, in order for HeS district legislators to be relatively representative of district-wide opinion, we expect that they have had time to nurture and stabilize a strong base of support which allows them to balance conflicting district demands. Second, if HeC district legislators are relatively junior, then accountability to party activists would be a likely explanation of the weak relationship between overall district preferences and roll-call votes. Available evidence suggests that junior legislators, much more than senior legislators, feel directly accountable to the party activists who have helped them get elected. Wright (1978; Fiorina, 1974:110-111) has argued convincingly, for example, that opponents of incumbents are likely to look to their party supporters in the primary election as their most crucial source of support, while incumbents are more likely to view their support in district-wide terms. As a logical extension it is reasonable to assume that legislators will need several terms to broaden their electoral base beyond the party activists who provided their initial support.

Table 6, which reports data on the seniority of legislators by type of district, upholds this argument. In both years legislators from HeS districts are discernibly more senior than HeC district representatives. Perhaps the more telling of the two statistics reported in Table 6 is the percentage of representatives who had served two terms or less. In 1971 less than 4% of HeS legislators are junior; in 1973 16% had served two or less terms. In contrast one-third of the 1971 and more than 60% of the 1973 HeC district representatives are junior.

Although the evidence is indirect, it nonetheless substantiates our contention that HeC and HeS district legislators respond in different ways to their heterogeneous constituencies.[13] If our thesis is correct, we also have identified a situation in which the response by legislators to their constituencies is

TABLE 7
SENIORITY OF LEGISLATORS BY TYPE OF HOMOGENEOUS DISTRICT

Seniority	1971 HoC	1971 HoS	1973 HoC	1973 HoS
Median Years Served	5.5	6.4	6.0	6.6
Percent of Representatives Serving Two or Less Terms	30%	11%	25%	17%

not reflected in the overall level of policy agreement. In fact, since legislators from heterogeneous districts by definition face greater conflict, and thus higher uncertainty, it is probable that they are *more* responsive than representatives from homogeneous districts. The nature of the district, however, precludes a high level of overall policy agreement. This is particularly true of the HeC legislator, who has not yet had time to broaden his electoral base, and consequently is directly accountable to an active and ideological group within the district.

Implicit in the preceding discussion is an assumption that representatives may determine, to some extent, their own electoral fortunes by voting optimally (see Fiorina, 1974). It underlies, for example, our discussion of efforts by HeS legislators to nurture and stabilize an electoral base over time. The question then raised, however, is why don't HoC district representatives fare better? Indeed, optimal voting should have the greatest payoff for legislators who represent constituencies characterized by a consensus of opinion. One possible answer, supported by the data, is that representatives from HoC districts tend to be Republicans who were victorious in predominately Democratic districts, or, to a lesser extent, Democrats who won in predominately Republican districts.[14] Under these circumstances, even though representatives vote optimally, i.e., they respond primarily and consistently to visible majority preferences, they remain in a competitive situation because of the distribution of party loyalties among the electorate.

This is not to suggest that their efforts to maximize their roll-call votes is without *any* reward. Table 7 reports information on the seniority of legislators from homogeneous districts. The data indicate that legislators from HoC districts, even though they continually face competitive races, are able to keep themselves in office. Among the 1971 HoC legislators, 70% had served more than two terms. Similarly 65% of the 1973 HoC district legislators are not junior. Apparently, a legislator's judicious use of roll-call votes does not go completely unnoticed within his district.

We have summarized our discussion regarding the responsiveness of legislators from the various types of districts in Table 8. Again we emphasize the discrepancy between the efforts which legislators make on the one hand and

TABLE 8
A SUGGESTED PATTERN OF REPRESENTATIVE BEHAVIOR, BY DISTRICT TYPE

Type of District	Level of Policy Agreement	Extent of Responsiveness by Legislator	To Whom Accountable
HoC	High	High	Majority
HoS	Moderate	Moderate	Majority
HeC	Moderate/Low	High	Party Activists
HoS	Moderate/High	High	Coalition of Groups

the resulting level of policy agreement on the other. Responsiveness does not necessarily imply policy agreement. In terms of the data normally available to researchers, regression coefficients can measure the latter, but not the former.

Our findings also have implications for the representation of minority preferences. Most evident is the fact that minority preferences are likely to receive little attention in HoC districts. HoC district representatives are responsive, but predominately to the dictates of the majority. HoS district legislators are somewhat more likely to eschew majority preferences on occasion and thus are more likely to respond to minority interests. On very salient policy questions, however, we would not expect HoS district representatives to ignore majority sentiments. A case in point is the roll call behavior of southern congressmen on civil rights issues during the 1950s. To the extent minority preferences receive attention, then, it is most likely to occur in heterogeneous districts. Given the configurations within heterogeneous districts, representatives may feel that they can't afford not to respond to minority interests at least some of the time.

CONCLUSION

Representative government is a derivative form of government; it is, as *Federalist Paper No. 52* (Hamilton, 1961:327) notes, "a substitute for the meeting of citizens in person. . . . ." It is only natural, therefore, that a central concern of representative government should be the institutionalization of mechanisms which lead the few who govern to be accountable to the many who are governed.

Our purpose has been to ascertain whether the heightened prospect of electoral accountability, as measured by district marginality, encourages more responsive behavior on the part of representatives. To the extent our data allow an answer, it is affirmative. But it is also a complex answer, made so in large part because the homogeneous/heterogeneous nature of the district looms large in determining to whom the legislator will respond. In homogeneous districts small margins apparently cause representatives to feel more accountable to the overwhelming majority than legislators who are electorally safe, as the marginality hypothesis predicts. But when the configurations within a district are complex, as they typically are in heterogeneous districts, small electoral margins tend, perhaps paradoxically, to both heighten responsiveness and reduce the over-all level of policy agreement. Being in a situation of high electoral accountability *and* suffering from mandate uncertainty, the legislator (typically junior) from a district which is both competitive and heterogeneous apparently remains, in the short run, accountable primarily to the party activists who originally helped him gain office. The end result, at least as measured here, is that the over-all level of policy agreement is weak.

To say that small electoral margins lead to heightened responsiveness is not to say that electoral safety also leads to unresponsiveness. Although representatives from safe districts which are also homogeneous evince a tendency to ignore district sentiment, their flexibility is unquestionably limited. Not responding to majority sentiments on the key policy questions may be equivalent to committing electoral suicide. Similarly, it appears that representatives from heterogeneous districts can improve their electoral fortunes by carving out, over time, a reasonably dependable electoral majority. But they then become directly accountable to them.

Our exploration indicates, then, that the increased probability of electoral accountability does enhance the responsiveness of representatives, but often in ways which belie the simplicity of the marginality hypothesis, at least as it has been stated by researchers. But the nature of our data has necessitated speculation. If these speculations encourage additional research, based on data which more directly get at the intricate and dynamic processes which occur between representative and represented, our purpose will have been served.

NOTES

1. Prewitt (1970) has challenged the validity of this argument by showing that many local legislators have no intention of seeking reelection. For these legislators, he argues, electoral accountability itself cannot ensure responsiveness.

2. The recent proliferation of research on the vanishing marginals of incumbent congressmen (Kostroski, 1973; Mayhew, 1974; Ferejohn, 1977; Fiorina, 1977; Cover, 1977) stems in part from the concern that increased electoral security will allow incumbents to ignore the political sentiments of the public.

3. A related stream of research has examined whether legislators from marginal districts display more moderation in their voting behavior. Sullivan and Uslaner (1977) provide a good summary of this literature.

4. A notable exception is Deckard's (1976) analysis of the 88th and 92nd Con-

gresses. She shows that marginality has no independent effect when district typicality is controlled.

5. Scholars increasingly are attacking this measure because of the small within-district samples. See Erikson (1977) for an excellent discussion of the sampling problems related to the SRC study.

6. When Fiorina adds a dummy variable for incumbency, the standard error drops to a still rather substantial .12.

7. We can also use this scheme to measure the independent effect of district homogeneity, controlling for marginality. The relevant comparisons then are A-C and B-D. Although important, we shall not address this concern here.

8. Other studies which rely upon referenda returns include Crane (1960), Hedlund and Friesema (1972), and Erikson et al. (1975).

9. It is important to note that we are attempting to identify the amount of consensus on the dimension as a whole and not the individual issues. Thus districts with different combinations of proposition voting may be classified as equally homogeneous or heterogeneous. Consider, for example, a district in which the percent vote on five related propositions is 50, 50, 50, 70, 30. This district would be defined as maximally heterogeneous. But so would the district whose percent vote on the five propositions is 30, 30, 70, 70, 50. If the representational linkage occurs through broad policy dimensions, then this measure ought to fit the calculus of a legislator who is attempting to assess consensus within his district on a particular domain.

It is also important to caution that we are talking about homogeneity only in terms of the distribution of district opinion. Other scholars define heterogeneity in terms of the structural differentiation of the district, i.e., its population attributes and/or ecological characteristics. The distinction between these two conceptions may be important to the findings reported below.

10. Roll-call behavior was measured in the session immediately following the election.

11. The advantage of using all observations is that the estimates will be more efficient. When using a single subset, the estimate of the population variance, δ^2, does not utilize the information about the δ^2's, contained in the remaining subsets (Kmenta, 1971:421). In using a single equation, we have assumed that the variance of the error term is the same for all subgroups.

12. Fiorina (1974) has developed a theoretical model to predict the behavior of representatives under different constituency conditions.

13. An analysis of the residuals also suggests that junior legislators are responding primarily to active members of their parties. The residuals indicate that the HeC district legislators who are also Democrats are clearly more liberal than is predicted by their constituency opinion scores. Similarly, Republicans vote more conservatively than district preferences predict. The residuals are discernibly smaller among HeS district legislators.

14. Legislators from HoS districts, in contrast, very rarely find themselves representing districts characterized by an unfavorable distribution of party loyalties. Similarly, the distribution of party loyalties does little to explain why HeC districts are competitive.

REFERENCES

CLAUSEN, A. R. (1973). How congressmen decide: A policy focus. New York: St. Martin's.

COVER, A. D. (1977). "One good term deserves another: The advantage of incumbency in congressional elections." American Journal of Political Science, 21(August):523-541.

CRANE, W. (1960). "Do representatives represent?" Journal of Politics (May):295-299.

DAHL, R. A. (1956). A preface to democratic theory. Chicago: University of Chicago Press.

DECKARD, B. S. (1976). "Electoral marginality and party loyalty in House roll-call voting." American Journal of Political Science, 20(August):469-481.

DYE, T. R. (1961). "A comparison of constituency influences in the upper and lower chambers of a state legislature." Western Political Quarterly, 14(June):473-480.

ERIKSON, R. S. (1977). "Constituency opinion and congressional behavior: A re-examination of the Miller-Stokes representation data." Unpublished manuscript.

ERIKSON, R. S., LUTTBEG, N. R., and HOLLOWAY, W. V. (1975). "Knowing one's district: How legislators predict referendum voting." American Journal of Political Science, 19(May):231-246.

EULAU, H., and KARPS, P. D. (1977). "The puzzle of representation: Specifying components of responsiveness." Legislative Studies Quarterly, 2(August):233-254.

FEREJOHN, J. A. (1977). "On the decline of competition in congressional elections." American Political Science Review, 71(March):166-176.

FIORINA, M. P. (1977). "The case of the vanishing marginals: The bureaucracy did it." American Political Science Review, 71(March):177-181.

--- (1974). Representatives, roll calls, and constituencies. Lexington, Mass.: D. C. Heath.

--- (1973). "Electoral margins, constituency influence, and policy moderation: A critical assessment." American Politics Quarterly, 1(October):479-498.

HAMILTON, A., et al. (1961). Federalist paper No. 52. p. 327 in C. Rossiter (ed.), The Federalist papers. New York: Mentor.

HEDLUND, R. D., and FREISEMA, H. P. (1972). "Representatives' perceptions of constituency opinion." Journal of Politics, 34(August):730-752.

JANDA, K. (1961). "Democratic theory and legislative behavior: A study of legislator-constituency relationships." Unpublished Ph.D. thesis, Indiana University, Bloomington, Indiana.

JEWELL, M. E., and PATTERSON, S. C. (1977). The legislative process in the United States (3rd ed.). New York: Random House.

JONES, B. D. (1973). "Competitiveness, role orientations, and legislative responsiveness." Journal of Politics, 35(November):924-947.

KMENTA, J. (1971). Elements of econometrics. New York: MacMillan.

KOSTROSKI, W. L. (1973). Party and incumbency in post-war Senate elections: Trends, patterns, and models." American Political Science Review, 68(December):1213-1234.

KUKLINSKI, J. H. (1978). "Representativeness and elections: A policy analysis." American Political Science Review, 72(March):165-177.

--- (1977). "District competitiveness and legislative roll-call behavior: A reassessment of the marginality hypothesis." American Journal of Political Science, 21(August):627-638.

Le BLANC, H. (1969). "Voting in state senates: Party and constituency influences." Midwest Journal of Political Science, 13(February):33-57.

LUTTBEG, N. R. (1974). Public opinion and public policy (rev. ed.). Homewood, Ill.: Dorsey.

MacRAE, D., Jr. (1952). "The relation between roll calls and constituencies in the Massachusetts House of Representatives." American Political Science Review, 46(December):1046-1055.

MAYHEW, D. (1974). "Congressional elections: The case of the vanishing marginals." Polity, 6(Spring):295-317.

MILLER, J., and ERICKSON, M. (1974). "On doing dummy variable regression analysis: A description and illustration of the method." Sociological Methods and Research, 2(May):409-430.

MILLER, W. E. (1970). "Majority rule and the representative system of government." Pp. 284-311 in E. Allardt and S. Rokkan (eds.), Mass politics. New York: Free Press.

MILLER, W. E., and STOKES, D. E. (1963). "Constituency influence in Congress." American Political Science Review, 57(March):45-56.

PATTERSON, S. C. (1961). "The role of the deviant in the state legislative system: The Wisconsin Assembly." Western Political Quarterly, 14(June):460-472.

PESONEN, P. (1963). "Close and safe state elections in Massachusetts." Midwest Journal of Political Science, 7(February):54-70.

PREWITT, K. (1970). "Political ambitions, volunteerism, and electoral accountability." American Political Science Review, 64(March):5-17.

SCHUMPETER, J. A. (1947). Capitalism, socialism and democracy. New York: Harper & Row.

SHANNON, W. (1968). "Electoral margins and voting behavior in the House of Representatives: The case of the Eighty-Sixth and Eighty-Seventh Congresses." Journal of Politics, 39(November):1028-1045.

STONE, W. J. (1977). "A panel analysis of representation in Congress: A preliminary report." Paper presented at the annual meeting of the American Political Science Association, Washington, D.C., September 4.

SULLIVAN, J. L., and USLANER, E. M. (1977). "Congressional behavior and electoral marginality." Unpublished manuscript.

WRIGHT, G. C., Jr. (1978). "Issue strategy in congressional elections: The impact of the primary electorate." Paper presented at the annual meeting of the Midwest Political Science Association, Chicago, Illinois, April 20-22.

10

Interest Groups and Accountability: An Incompatible Pair

MICHAEL A. BAER

□ ACCOUNTABILITY IMPLIES LINKAGES. We use the concept of accountability in a variety of situations. It may be used in discussing the responsiveness of a public agent or agency to elected officials. That is, we ask how closely an agency carries out the intentions of the state legislature when a state law is being implemented. It may also be used in considering the linkage between the elected official and the voting public. A public that is aware of the actions of elected public officials and of electoral procedures may try to influence the actions of these officials. In this way the public may hold the official accountable for his actions. Conversely, in this situation, the public official may be placed in the position of justifying his actions to the public.

Interest groups, too, play a linkage role in our political system. They are representatives, though, of specific interests, rather than geographical districts. However, even though interest groups are representatives, to use the concept *accountability* in the same breath as *interest groups* raises the question of compatibility of terms. If interest groups can hold public officials or governmental agencies accountable, we must ask the question: accountable to whom? Is the agency being held accountable to the public, to a specific segment of the public, or only to group leadership? We must also ask if we can clearly separate responsiveness from accountability. All public agencies should be accountable for their actions to all segments of the society. Interest groups and individuals of all persuasions can expect public agencies and elected officials to make public their actions and to be able to justify them.

But not everyone can expect public agencies to be responsive to their own desires or preferences.

As the following example illustrates, we must be able to distinguish between accountability and responsiveness: I am reminded of a small town bisected by a river. A highway connecting two major population centers runs through the town. There is an acceptably wide roadway through the town on either side of the river, but, in both cases, the roadway winds through residential neighborhoods. As might be expected, neighborhood associations from each side of the river try to persuade the county council to route the highway traffic onto the "other side" of the river. To whom is the council accountable? To whom should it be responsive? Can and should each of these neighborhood associations hold the council accountable? Does accountability only imply that the council must explain why it takes the actions it does? Obviously, it cannot be responsive to both groups (without constructing a third road), but it can be accountable to both.

There are some questions (many raised by this simple example) that must be discussed when we tie accountability and interest groups. In fact, if accountability is used to express anything other than making public an action and justifying that action, then the concept of an interest group *holding* the government accountable is antithetical to that of elected representative government. As accountability often seems to be used to mean more, the term used in this way is not compatible with interest groups.

As we are aware, interest groups have the capacity to influence the governmental process at several stages. From the possibility of agenda setting at the pre-legislative stage through post-legislative influence on policy implementation, the impact of interest groups can certainly be observed. An inventory of the points of input that interest groups may utilize would, at a minimum, include influencing public opinion through media usage, participation in the recruitment and election of candidates, influence in the selection of appointed officials, participation in the legislative process, indirect and direct impact on regulatory and administrative agencies, and impact on interpretation and enforcement of law through litigation in the courts.

At each of these stages there is the possibility of interest groups not only influencing the political participants in their decisions, but also requiring them to justify their actions. Thus we could say that interest groups acting for their own interests, and as representatives in a pluralistic society, push public agencies toward accountability by forcing communication. Though once again we must note the difference in the use of accountability here and the way it might be used by a legislative oversight committee.

The methods used by interest groups to influence public policy vary with the groups themselves. There are a variety of types of groups, each having grown in different ways, each having different purposes and goals. These range from the minimally organized to those well established, well financed groups which have been in existence for many years. Examples of the less organized include relatively spontaneous citizen groups banding together to protest controversial single issues such as placement of a highway, building of nuclear reactors, or the prosecution of individuals on what is perceived to be political grounds. These less organized groups also often include neighborhood associations which frequently lay dormant until a local controversial issue arises. Among the more stable, longer tenured groups, we can include those groups which serve private interests such as business associations, industry representatives, and professional and occupational groups (such as unions or educational lobbies). Related, but in a slightly different category, are federations of local governments, such as municipal leagues, which offer service, research, and lobby activities to their members.

We can also find groups which claim to represent the public interest such as consumer organizations, nonpartisan groups as the League of Women Voters, and more recently groups, such as Common Cause, aspiring to act as watchdog for the citizenry over a broad range of issues, including governmental reform. Finally, we might also include the numerous governmental advisory committees, particularly those appointed to advise regulatory agencies. Because of their history and selection processes often employed in appointing members, it is difficult to know whether these quasi-governmental bodies represent public or private interests (Turner, 1972). As each of these types of groups have different reasons for their growth, we would expect their members (both individual and organizational) to have different expectations for the activities and the services to be provided by the groups.

Since it is a combination of who these groups represent, how they grew, and what their purposes are that influence their needs and abilities, and thus their efforts to hold governmental units accountable, let us examine each of these types of groups. Olson has asserted that most "large economic groups with significant lobbying organizations . . . are also organized for some other purpose." (Olson, 1967:132) That is, Olson does not think that influencing and holding governmental units accountable would be enough of a service to members to warrant the continued existence of most of these organizations. Marsh's examination of the Confederation of British Industries (Marsh, 1976) and Browne's study of municipal associations in the St. Louis area (Browne, 1976) would tend to confirm this assertion of Olson's.

Yet, though the membership may not consider lobbying the primary reason for joining a business association, a governmental group, or a labor union, the fact is these organizations do participate in shaping public policy. They do influence legislation and their presence in the political arena has an effect on the implementation of these legislative acts.

When one examines these large economic interests, the question of who they actually represent when they present policy alternatives to government becomes an intriguing question. One pictures the large business association representing the collective interests of the industry. However, it is often the case that these associations present opinions on issues much broader than that which directly concerns the particular industry.

Similarly, the interests of labor unions could not be monotonic given the varied interests of the numerous members of the union. In most instances there are electoral procedures within an organization that set up governing boards to whom lobbyists themselves are accountable. However, as Luttbeg and Zeigler (1966) have demonstrated in their study of the relationship between educators in Oregon and the group leaders representing the educators, there are often discrepancies between views of the membership and the representation of the group leadership.

It is not at all certain that all members, either individual or organizational, of a group which tries to influence government, wish to be involved in the policy-making aspects. Browne (1976:264) noted that the mayor of a member municipality to a local government federation stated that while he could not deny the importance of lobbying, he also had no interest in it. The mayor claimed to know nothing about lobbying and indicated he would rely upon the federation to communicate with state and federal decision makers for him.

While members of the groups Browne studied did not choose to have direct input, and the staff lobbyists were more or less free agents, the members when asked did heavily approve of the legislative efforts of the municipal associations, even in several controversial areas. In other words, while the lobbyists were free agents, they recognized their accountability to the member cities well enough to stay within acceptable bounds on a variety of controversial issues.

While a variety of overt techniques may be utilized to persuade legislators and administrators (Zeigler and Baer, 1969:162ff; Milbrath, 1963), simply being there may be one of the main means by which interest groups hold government accountable. The fact that a group with a known position exists and will communicate dissatisfaction with policy to government officials and agencies acts as an incentive to have the agencies stay within certain boundaries when implementing policy. This is one of the arguments frequently mentioned for the establishment of government-supported consumer affairs offices. Such offices provide not only the ability for communication and complaint, but also act as a center of awareness and a public oversight institution.

Many of the recently formed public interest groups have also indicated that communication, oversight, and accountability to the public is their main raison d'être. These public interest groups are often established by citizens who have seen both implicit and explicit influence exerted by private interests and have organized in order to place pressure on behalf of the "public interest" upon government. These groups often take the position that government has been responsive to private interests to the detriment of the public interest. Thus the public interest groups proclaim that they will act as a counter influence on the government and hold it accountable for the public.

This raises the question of who the public interest groups really represent and is there such a thing as the public interest. Those who speak for public interest groups often claim to have found the answer to Schattschneider's (1975) question, "How can people get control of the government?" While Schattschneider postulated that interest groups "sing(s) with a strong upper class accent," not all public interest groups do. However, most do have an accent that can be connected with specific goals which makes it impossible for them to represent a wide and varied public. Public interests can only press for public responsiveness if a monotonic public interest exists. We can ask again, what is the public interest in the case of our bisected town?

Public interest groups can and do provide further communication between segments of the population and the government. They can increase the necessity for officials and agencies to communicate with the public, and they can, as do private interests, influence the government. However, it is more difficult to objectively assess their interests as the public interest. Thus just as we might reject the concept that government should be responsive to private interests, we may also reject the concept of government being responsive to "public interest" groups.

In addition to the legislative and administrative lobbying tactics used by many private interests, there is another tactic that is particularly relied upon by the public interests. That technique is litigation in the courts. A recent example of how such litigation has been used can be found in the recent reform of political campaign finances.

In 1971 the Congress passed the Federal Election Campaign Act. Common Cause developed a litigative strategy in insuring that these laws were interpreted and upheld as the group had envisaged them. Between 1971 and 1975 Common Cause filed seven cases in the Federal District Court of the District of Columbia to ensure that existing provisions of election laws were enforced. These cases dealt with violations of campaign contribution and expenditure laws. All of these cases were selected so that they emphasized not only the broad goal of ensuring compliance, but they were also selected so as to point out vulnerable points in the laws that would force Congress to further deal with these "weaknesses." To raise the funds that litigation requires, Common Cause also went the route of publicizing their actions and producing convincing research to present their case to the public through the media. Thus they were combining litigative, legislative, and public opinion activities to force public agencies to justify their actions (Fleishman and Greenwald, 1976). In this sense they were in fact holding agencies accountable for their actions. Common Cause was certainly not the first interest group to use this combination of tactics when their goals could not be achieved solely through the legislative arena (Ziegler, 1964:300ff.). The NAACP, for example, had previously been very successful using these techniques when they were trying to desegregate housing (Vose, 1967).

Another type of group activity is direct citizen protests. As we noted earlier, such protests are most often over single issues. Let us take the example of a citizen's group who might be protesting the location of an airport in a metropolitan area. The protests of such a group may be centered around general environmental issues, the destruction of a neighborhood because of noise, because of added traffic, or because of roadways that would slice a neighborhood into sections. They may include large numbers, even the majority of the residents within the area of the city concerned. A variety of tactics may be employed from court cases, administrative appeals, electoral threats to incumbents, petitions or representations to local officials, and so on. However, the results they are seeking are responsiveness to the groups' own perception of local interests. Only indirectly do these actions lead to accountability. The need for the local planning board, the city council, or administrators to explain the rationale for their decision is often forced by such protest activity. Such activity furthers the communication between officials and the citizenry, and thus causes the decision makers to be accountable for their actions whatever the final decision is.

The discussion in this chapter has presented a conflict between the notion of accountability and responsiveness. It has noted that there are several links in our political system where both concepts are equated. An example of this is the legislator who is accountable to the public and thus is generally responsive to public desires. Another is the agency that is accountable to the legislature, responding to legislative directions. But when we enter into a discussion of interest groups the terms accountability and responsiveness diverge. Certainly interest groups provide a means of communication between segments of the society and legislators, administrators and other policy makers. These communication links are an important part of a pluralistic society. The communication, both formal and informal, forces the agencies to make public their decisions, to justify their actions, but in no way need agencies try to always respond to the policies favored by interest groups. Thus accountability need not imply responsiveness.

REFERENCES

FLEISHMAN, J. L. and GREENWALD, C. S. (1976). "Public interest litigation and political finance reform." Annals, 425 (May):114-123.

LUTTBEG, N. R. and ZEIGLER, H. (1966). "Attitude consensus and conflict in an interest group: An assessment of cohesion." American Political Science Review, 60 (September):655-665.

MARSH, D. (1976). "On joining interest groups: An empirical consideration of the work of Mancur Olson, Jr." British Journal of Political Science, 6(April):257-271.

MILBRATH, L. (1963). The Washington lobbyists. Chicago: Rand McNally.

OLSON, M., Jr. (1967) The logic of collective action: Public goods and the theory of groups. Cambridge, Mass.: Harvard University Press.

SCHATTSCHNEIDER, E. E. (1975). The semisovereign people (2nd ed.). Hinsdale, Ill.: Dryden Press.

TURNER, E. W. (1972). "Advisory committees: The fifth branch of government." The Bureaucrat (Summer):142-149.

VOSE, C. (1967). Caucasians only. Berkeley: University of California Press.

ZEIGLER, H. (1964). Interest groups in American society. Englewood Cliffs, N.J.: Prentice-Hall.

ZEIGLER, H. and BAER, M. A. (1969). Lobbying: Interaction and influence in American state legislatures. Belmont, Cal.: Wadsworth Publishing.

11

Fiscal Accountability: The Role of the Auditors

FLOYD E. STONER

□ TRADITIONALLY, FISCAL ACCOUNTABILITY meant that administrators responsible for public funds were required to maintain accurate accounts of expenditures that could be investigated by independent agents. Systems of state audit provided the means of implementing this narrow conception of fiscal accountability for thousands of years (Normanton, 1966). The focus of such systems was upon graft and corruption, and their investigations extended no further than the honesty of administrators. Other aspects of what has more recently been emphasized concerning accountability (e.g., competence, creativity, programs' impacts) was of no concern to the public auditors. This situation has changed dramatically. Governmental auditors in the United States currently claim to be the primary agents of accountability, broadly defined.

Basing their activities squarely on legitimacy derived from the traditional conception of fiscal accountability, some auditors today are investigating program administrators' compliance with laws and regulations, their efficient and economical management of resources, and the programs' performance (Staats, 1975:64). In the last 30 years, especially at the federal level, the auditors have been major exponents of this expansion of their role. While these changes are most visible at the federal level, the new conception of audits is influencing state government as well (Dittenhofer, 1970; Knighton, 1970; Graham, 1970).

AUTHOR'S NOTE: *I wish to thank Kenneth Farmer, Brett Hawkins, Peter Eisinger, and the editors of this volume for their helpful comments on an earlier version of this paper.*

Of what importance to policy analysts is this expansion of the auditors' role? It is important because as the auditors enlarge the scope of their investigations, they necessarily redefine the concept of fiscal accountability for those charged with the expenditure of public funds. No longer is "mere" honesty acceptable. This new approach to fiscal accountability requires efficient and economical management of resources to achieve the maximum impact in accord with legislative intent. This last phrase, "in accord with legislative intent," constitutes one of the most problematic areas of the auditors' expanded role. Who shall define the intent of legislators so that program administrators may be held accountable to it? The issue of legislative intent cannot be ignored in a discussion of fiscal accountability, because it is central to the current operationalization of the concept by the auditors.

This article presents an investigation of the relationship between fiscal accountability and public audits, with an emphasis on the issue of legislative intent.[1] The relationship is important. The auditors may be impartial in their expanded audit coverage, but the results of their audits are not neutral. The article is broken into three parts. First, the case for an expanded role for auditors is discussed. Second, the system of auditing overseen by the General Accounting Office (GAO) is considered. The last section is a case study of the impact of the auditors on the implementation of one particular policy, Title I of the Elementary and Secondary Education Act. As the case study illustrates, auditors' investigations and reports constitute a new input to the policy process, and an input of rapidly growing importance.

FISCAL ACCOUNTABILITY AND PUBLIC AUDIT

Audits by legislative auditors are the primary means of guaranteeing the accountability of the executive branch to the legislative branch in a system of government with a separation of powers. The budget is the tool for legislative control, and the legislature's auditors provide the information which can be used by the legislators to determine whether their wishes, expressed in the budget, have been carried out (Normanton, 1966:1-8). As long as government was limited in size and governmental operations were carried out under fairly precise delegations of legislative authority, the duty of the auditors was rather simple. Basically, they checked the books of government agencies to determine whether spending was within the categories specified in the budget and within the spending limits. This information was then provided to the legislators. Fiscal accountability, thus understood, was very limited and related directly to spending.

The problem with this traditional approach, according to some (Staats, 1975; Normanton, 1966) who would enlarge the scope of audits, is that the information produced is of little utility to modern decision makers in complex society. A final financial account provides no explanations for its "headings and totals." There is no way of telling, from a financial record, what administrative behaviors were actually financed by public funds. The money may have been spent legally, but was it spent economically and efficiently on activities and materials with a high probability of achieving the goal established in the legislation? Those with a broader view of the auditing function argue that in complex society audits must be designed to answer such questions so that these post hoc accountability investigations of public expenditures can inform the planning of future modifications of programs (Normanton, 1966:24).

Following this line of reasoning, those who support the new role for auditors argue that a complete audit program should have several features. There is general agreement that audits should cover traditional fiscal concerns, check compliance with laws and regulations, investigate management practices, and include some measures of the effectiveness of programs (Dittenhofer, 1970:181; Staats, 1975:64). There is one point of overriding importance that must be made concerning this approach to auditing, and it is related to the performance of such audits. While the argument may seem extremely compelling (who can oppose the desire of auditors to provide more complete information to decision-makers),[2] the problem is that acceptance of such a conception of the auditing function necessarily involves a major change in the activities of the auditors. The implications of such changes are not immediately apparent.

Checking compliance with laws and regulations at the state and local level, evaluating management practices on-site, and assessing program performance necessarily involves the auditors in determinations of legislative intent. How else can they make sense of the masses of regulations; how else can they determine whether a program is achieving goals established by Congress? While the auditors can argue that they are only providing information to decision makers, such a public investigation of all aspects of administration can have a major impact on the conduct of the administrators. When the investigation is tied to the auditors' determination of legislative intent, the auditors can greatly influence the operational definition of that intent at the state and local level by means of their audits. While in theory it is obvious that administrators are responsible for all of their official acts, the expansion of public audits to cover all such activities may shift the focus of power concerning accountability from the Congress to the audit office. The question then becomes, who holds the agents of accountability accountable?

FEDERAL SYSTEM OF STATE AUDIT

The federal system of state audit was first established by the Budget and Accounting Act of 1921. That legislation created the GAO as a congressional support agency with the charge to investigate "all matters relating to the receipt, disbursement, and application of public funds" (quoted in McMickle and Elrod, 1974:13). The GAO at first concentrated upon "routine examination of a staggering volume of vouchers." (Fitzgerald, 1975:384). These examinations were performed in Washington and dealt with the issue of regularity of expenditures.

After World War II, the GAO underwent a major transformation. Under prodding by the Hoover Commission and a new Comptroller General, the GAO began a steady expansion of activities by developing the "comprehensive audit" concept. The Comptroller General (quoted in Normanton, 1966:113) described the purpose of a comprehensive audit in 1954:

> To determine to what extent the agency under audit has discharged its financial responsibilities, which imply equally the expenditure of public funds and the utilization of materials and personnel, within the limit of its programmes and activities and their execution in an effective, efficient and economical fashion.

The move away from the traditional conception of fiscal accountability began with this management emphasis, despite the fact that the Budget and Accounting Procedures Act of 1950 "reflected a perspective that was entirely financial in character" (Pois, 1975:248). According to Joseph Pois (1975:248) this push for "broad-based reviews . . . did not come from the language of the Office's statutory charter, but from within the GAO itself."

The pressure was successful. "The Legislative Reorganization Act of 1970 for the first time provided a specific statutory basis for the type of broader review such as the GAO had already been conducting" (Pois, 1975:249). By 1970, without giving up its position as the primary agent of traditional fiscal accountability, the GAO had also acquired statutory authority for investigations of management practices and the achievement of program results.

What do these changes in the statutes mean? An insightful article by Ira Sharkansky (1975) raises most of the important issues surrounding the GAO. At base, all of the issues are related to the ambiguity of the auditors' job. When Congress enlarged the GAO's mandate, it increased the ambiguity geometrically because it gave the auditors even more of a basis on which to establish for themselves what activities were proper to their role.

The definitive statement of what fiscal accountability means is contained in the *GAO's Standards for Audit of Governmental Organizations, Programs, Activities and Functions* (U.S. Comptroller General, 1972). By establishing standards for audits of any governmental "organizations, programs, activities and functions" no matter who performs them (U.S. Comptroller General, 1972:1), the GAO operationally defines fiscal accountability for both legislators and program administrators. The scope of that accountability is indicated by the following summary of the specifics of the GAO's "comprehensive audit" contained in the *HEW Audit Agency Policy Handbook* (1974:17):

1. The full scope of an audit of a governmental program, function, activity, or organization should encompass:
   a. An examination of financial transactions, accounts, and reports, including an evaluation of compliance with applicable laws and regulations.
   b. A review of efficiency and economy in the use of resources.
   c. A review to determine whether desired results are effectively achieved.

For the auditors, these points are the foci of their investigations; for program administrators, these points mean that any aspect of administrative behavior may be scrutinized and reported upon.

By what standard does the GAO claim that administrators should be judged? As noted earlier, expansion of the auditors' role is leading to ever greater centralization of power in the auditors' hands. This issue of standards on which to base audits is a crucial example. According to the GAO's *Standards,* any audit must begin with a determination of congressional intent.[3] Governmental auditors are directed to examine the laws, regulations, committee hearings, committee reports, and any other program materials and reports relevant to the audit (U.S. Comptroller General, 1972:28-29). On the basis of this material, auditors are expected to establish the "purpose" of the audited entity so they can use that "intent" as the basic standard for their audits, whether the audits emphasize compliance, management or performance. For both auditors and administrators, the concept of fiscal accountability has undergone a major expansion since World War II.

These changes in the meaning of "fiscal accountability" have been pushed by members of the GAO. What makes them even more important is the fact that the system of audits overseen by the GAO extends deep into the workings of the federal government. The changes initiated by the GAO have influenced the entire system.

Section 117(a) of the Budget and Accounting Act of 1950 gave the GAO responsibility for "the effectiveness of accounting organizations and systems, internal audit, and related administrative practices of the respective agencies." Throughout the years since receiving that charge the GAO has gradually increased the level of its involvement in the 50 internal audit organizations within federal departments and agencies. The GAO's involvement has been primarily in the form of audit standards and procedures and investigations of their implementation. The *Standards for Audit* were intended by the GAO to apply to these audit organizations, and in 1973 Federal Management Circular 73-2 was issued by the Office of Federal Management Policy, officially adopting the *Standards* for use by the internal audit agencies. In 1974 the GAO issued *Internal Auditing in Federal Agencies: Basic Principles, Standards and Concepts,* a publication that basically encouraged the extension of "comprehensive audits" to these audit agencies. Then, in 1976, the Comptroller General issued a report to Congress entitled "An Overview of Federal Internal Audit," which was based upon an investigation of these agencies. This report stressed the need to expand these agencies in order to provide the audit coverage formalized in the *Standards.*

While the GAO has a responsibility to audit virtually all executive branch activities, these internal audit agencies deal primarily with the programs of their departments. It is the internal audit agencies that investigate the administration of programs on a regular basis. The point of this discussion of internal audit is to indicate the extent to which the expanded conception of government auditing *may* influence the accountability expected of program administrators. Obviously, much of the impact depends upon the receptivity of these internal auditors to the new directives. However, if the internal auditors do accept the GAO's approach to audits, the implications are most far-reaching because of the regular audits performed by internal audit agencies.

TITLE I: A CASE STUDY[4]

The implementation of Title I of the Elementary and Secondary Education Act provides an interesting illustration of the impact of auditors in the policy process. The central issue is the auditors' role in the determination of congressional intent. The 12-year history of Title I illustrates the political ramifications of the expanded conception of fiscal accountability operationalized by the auditors.

As was noted, the standard upon which all federal auditors are directed to base their investigations is congressional intent. Of direct relevance to this point, Theodore Lowi (1969) describes, and Murray Edelman (1964) explains, a phenomenon of increasing frequency in the United States: ambiguous and symbolic legislation resulting in varied and idiosyncratic administration of public policies that greatly benefit established interest groups. For Lowi, this system of "interest-group liberalism" is unfortunate because it means that elected representatives do not make decisions. For Murray Edelman, the situation is understandable because it enables these politicians to provide benefits to important political supporters while reassuring the mass public that their interests are protected and "good" is being done. Neither writer expects much change. However, the auditors have expanded their role in recent years, and their activities do not fit well with vague delegations of authority.

Title I has many of the characteristics cited above: ambiguous statutory language and much administrative discretion. However, some of the administrators have not performed in the manner expected. Rather than simply carrying out the desires of congressmen and interest group representatives for a low profile distribution of funds to the existing educational system, they have attempted to establish a national program with specific goals designed to reform that system. The attempt might have been inconsequential were it not for the auditors. The Department of Health, Education and Welfare's internal auditors, reorganized and given a new charter in 1965, constitute an obstruction to the smooth functioning of the system of "interest-group liberalism."

For Title I to conform to the model described by Lowi, there should be few clear standards formulated by administrators, and those that are established should benefit organized groups (Lowi, 1969:101-24). The legislation was ambiguous; the money could simply have been sent to the states. However, some Office of Education (OE) administrators developed federal regulations designed to redistribute both resources and political influence in local school districts. Once established, the auditors utilized them in their audits of state administration of the funds. Their audits were broader in scope than the fiscal audits performed in the Department of Health, Education, and Welfare before 1965. They performed compliance audits to determine whether states and localities had spent the federal funds on programs designed to achieve the goal established by Congress. Their primary sources for ascertaining the "congressional intent" were the Title I program requirements (regulations and program materials produced by OE) with which a number of influential congressmen disagreed. Neither state and local administrators nor many congressmen expected program people to be held accountable by this type of audit.

Title I of the Elementary and Secondary Education Act of 1965 authorized the most direct federal involvement in local public education in United States' history. It is also the most heavily funded title of the act, which provides funds to school districts in 95% of all counties in the country. The organizational hierarchy designated to administer this program parallels the federal structure with distinct national, state, and local administrative agencies. The language of the title emphasizes aid to educationally deprived children by funding programs formulated by Local Education Agencies (LEA) to overcome the particular handicaps of deprived children in individual localities. These programs must be approved by State Education Agencies (SEA). The Office of Education has general responsibility for implementing and administering this venture into educational federalism.

Although categorical aid for educationally deprived children in areas of concentrations of poverty is the purpose stated in ESEA Title I and reemphasized in the 1973 report by the National Advisory Council on Education of Disadvantaged Children, the legislative situation in which it passed was complex, and the statutory language unclear. One congressional staff person who was intimately involved with the bill in 1965 asserts that Title I was carefully designed to approximate general aid to education in the distribution of funds, while emphasizing the language of compensatory education to save congressmen from charges of "federal meddling in local education." This language was viewed by most congressional proponents of the bill as the means of passing a federal aid to education bill after 100 years of failure (Eidenberg and Morey, 1969:77-80; Munger and Fenno, 1961; Thomas, 1975:38). The other titles were also important, but Title I was the largest and the most important title (Murphy, 1973:162-163). The fundamental issue in 1965 and still today is the dichotomy between categorical and general aid (Thomas, 1975:38). The principle of federal aid to education has been established, but the form that it should take has been hotly debated (Thomas, 1975:76; *Congressional Record,* 1974).

The general aid/categorical aid conflict was not readily apparent during the drafting and passage of the Elementary and Secondary Education Act. President Lyndon Johnson established only one major requirement for the education bill that he wanted to sponsor. He asked Douglas Cater, Lawrence O'Brien, and Commissioner of Education Francis Keppel to devise "legislation that Congress would pass" (Murphy, 1973:164). The Democratic majority of the House Education and Labor Committee, the committee in which many previous pieces of legislation designed to provide general aid had died, expressed similar sentiments. Two individuals who were working in the House at the time later wrote "the Democratic membership of the education committee in the House (where the key battle was to be fought) by and large took the position that they would be able to support any legislation that carried the endorsement of the Administration and the principal interest groups" (Eidenberg and Morey, 1969:76). The bill's drafters had a relatively free hand, provided they could gain the support of the two primary interest groups, the National Education Association (NEA) and the National Catholic Welfare Conference (later called the United States Catholic Conference [NCWC]).

The NEA and the NCWC were important because they represented the primary stumbling blocks to the achievement of this long-sought goal: general aid to education.[5] The National Education Association had previously opposed any proposals that included funds for parochial schools; the National Catholic Welfare Council just as strenuously had opposed bills that would fund only public schools.[6]

There was one other problem area faced by the drafters of ESEA: the issue of Federal control. Support for local control of education had bedeviled general aid proposals since 1870 (Lee, 1949:165). However, if the two primary interest groups could be placated, the persons charged with devising legislation that would pass believed that they could succeed in passing it

(Eidenberg and Morey, 1969:79-80). It was an important issue, however, and it would have great relevance for the role of the auditors in the years that followed passage.

ESEA Title I was passed and President Johnson signed it into law on April 11, 1965. However, the means whereby this drive for federal aid to education was brought to fruition set the stage for years of conflict during implementation that were not expected by most of the congressmen influential in the passage of the bill.

The device utilized to gain the support of the NEA and the NCWC was the concept of "child benefit." While the Title I funds would obviously be spent in school systems, the policy statement emphasized the needs of "educationally deprived" children in "local educational agencies serving areas with concentrations of children from low-income families" (Public Law 89-10, Title I, Sec. 201).[7] The formula for the computation of local school district funding levels used the number of children from poor families and those receiving Aid to Families with Dependent Children. The emphasis on poor children was politically potent, both because of the publicized concern about poverty in 1965 and because the concept side-stepped the issue of parochiaid. Title I funds aided the child, wherever he attended school. In that sense it was categorical aid.

However, the hearings on ESEA held by both the House General Education Subcommittee and the Senate Education Subcommittee provide convincing evidence that the congressmen influential in the area of education policy believed that Title I was general aid under another name.[8] The hearings also contain evidence of the compromise between the two interest groups that resulted in support for the bill, but which would have interesting ramifications later. One key point in the bill was the provision that "special educational services and arrangements" would be provided by the Local Education Agency (LEA) for educationally deprived children in private schools (Public Law 89-10, Title I, Sec. 205. (2) (2)). This provision was what guaranteed the support of the NCWC. However, Carl Perkins, Chairman of the General Education Subcommittee, made it quite clear that he believed that the money provided to the public schools under Title I would not be restricted to certain uses, and Secretary of HEW Celebrezze (House Hearings, 1965:140) agreed:

> Chairman Perkins: Now could this legislation, particularly under Title I, actually increase the range of choices open to the local autonomy enjoyed under this proposal and actually create more freedom on the part of the local school administration?
>
> Secretary Celebrezze: That is absolutely right and that is one of the purposes of the act; to strengthen the local school system.

The Administration's testimony supported the belief that Title I would fund clearly defined special programs for the educationally disadvantaged in parochial schools, and that it would fund whatever the local educational decision makers wanted in public schools (House Hearings, 1965: 980-981).

The situation was the same in the Senate, though stated even more clearly. Senator Dominick elicited from Secretary Celebrezze agreement that Title I funds could be used for anything in public schools, including construction of gymnasiums and cafeterias. Then:

> Going back to this private-public school question, my understanding from your presentation and from a brief summary of the bill is that the general grants under Title I apply only to the public schools.
>
> Secretary Celebrezze: That is right.
>
> Senator Dominick: That the monies which are to be made available to the private schools come in the form of special assistance or special programs or library services or things like that?
>
> Secretary Celebrezze: To the pupils rather than to the schools, Senator. [Senate Hearings, 1965:499]

For the congressional proponents of Title I, the problem that the policy addressed was the "fiscal crisis" faced by public schools, particularly those with low property tax bases. Chairman Perkins (House Hearings, 1965:1962-1963) clearly stated this position during the hearings:

> The whole impact here is the concentration of low income families in the school districts that cannot provide adequate service for these youngsters. So, we follow the same guidelines that we set up in the aid to school districts "impacted" by Federal installations legislation all the way through. There is no Federal control. We just make sure that the funds get into the school districts where the impact actually exists.

The representatives of the two primary interest groups involved in the drafting process agreed. The goal of almost all concerned was legislation that would deliver dollars to financially strapped school systems. The symbol of the educationally deprived, impoverished child was the means to pass the bill. As Chairman Wayne Morse of the Senate Education Subcommittee said: "Let us face it. We are going into federal aid for elementary and secondary schools . . . through the back door" (*Congressional Record,* April 7, 1965:57313).

The formula by which the Title I funds were to be distributed to states and to school districts was precisely stated in the law. What was left undefined was the issue of control over the expenditure of funds *within* school districts. While the formula directed funds to districts with high concentrations of children from families below the poverty level, in 1965 the statutory language did not target the money to poor children within school districts. In fact, the original language stated that both "acquisition of equipment and where necessary the construction of school facilities" were allowable expenditures (Public Law 89-10, Sec. 205(a)(i)). It was true that the law called for "programs and projects," but as Secretary Celebrezze's previously cited response to Senator Dominick's question illustrated, many members of Congress believed (and were encouraged by the Administration to believe) that such language was only included to avoid the NCWC's opposition to the law. They were told that the special programs were for parochial schools, with "general grants" for public schools. Specific programs for poor children were not mentioned. This act was a first step to general aid in the opinion of most congressmen, and virtually all congressmen wanted it so administered. For that reason, the law was ambiguous. In the words of a Democratic member of the House committee:

> The bill itself was really categorical aids with the categories broad enough to resemble general aid. We made the categories so broad that the aid splattered over the face of the entire school system and could be called general aid in principle while politically remaining categorical aids. [Quoted in Eidenberg and Morey, 1969:90-91]

TITLE I, INTEREST-GROUP LIBERALISM, AND THE WAR ON POVERTY

At the time of its passage ESEA Title I was an excellent example of a policy produced by a system based on interest-group liberalism. Organized private groups were the primary influences on the drafting of the bill. The statutory language was extremely ambiguous (with the significant exception of the formula by which the funds were to be distributed at the local level). Discretion was granted to educational administrators at the federal, state and local levels. This administrative structure was an example of the "creative federalism" disparaged by Lowi because definitive national goals could not be sought and federal decisions would not be made by Congress.

In 1965, all of the attributes of interest group liberalism seemed to mesh in a single policy. Therefore, one would expect that federal administrators would simply send the Title I funds to the state for distribution to the school systems that qualified under the formula. One would expect few requirements to be attached to the use of the money at the local level. One would certainly not expect to find large sums of money spent by LEA's to be disallowed by federal auditors on the grounds that it had not been spent in accord with the "intent of Congress." The hearings and committee reports of 1965 indicate that the congressmen influential in education policy wanted the money spent as the state or local administrators desired. However, something happened to this "general aid" in the process of implementation.

A most fascinating account of the first four years of Title I was written by its first division program director, John Hughes, in collaboration with his wife, Anne, carefully delineated the means whereby a small group of administrators (whom they term "reformers") sought to develop the program in a direction that was not desired by congressional proponents, the major interest group representatives, or most of the influential administrators in the Office of Education. The Hughes detail an extended struggle between general aid supporters ("traditionalists") and advocates of Title I as the means "to modify the system to serve the special needs of the poor as a target population" (Hughes and Hughes, 1972:33).

The conflict between the general aid and the categorical aid proponents in OE was based on two differing conceptions of the "poverty problem" facing public education. If one believed that the problem of local schools was insufficient funding, then federal funds *without strings attached* was a reasonable answer. If, on the other hand, one believed that the "poverty problem" was both an unwillingness as well as an inability to deal with poor children in local schools, then a tightly controlled categorical grant program provided the only hope for change. Simply to give callous and incompetent administrators more money would only result in their failing to reach poor children at a higher level of expenditure.

The reformers in OE did not share the faith in state and local administrators that had been evidenced by the congressmen. Therefore, the first five years of implementation of Title I encompassed a series of attempts by these individuals to tighten the "basic criteria" pertaining to state approval of LEA applications for Title I funds. If defeated by congressional and interest group pressure, they tried again later and frequently succeeded. Over time, more and tighter requirements for LEA participation in Title I were included in the regulations, guidelines, and program memoranda that states were expected to monitor.

At the heart of the reformers' program was the desire to concentrate the Title I funds on a relatively small number of children in such a manner that these funds would supplement the state and local expenditures on the same children (Murphy, 1973:177-178). Local public school districts were required to submit applications for their Title I funds to their state administrators in which they presented the special programs that had been designed for the carefully identified educationally deprived children. These children were to be chosen from particular schools that were "targeted" because the number of children from low income families was higher than the district average. By 1971, after much conflict both in OE and between OE and Congress, these requirements had been firmly established.

Establishing the requirements was not an easy task for the reform-oriented administrators. However, achieving compliance by local administrators would prove to be even more difficult. In fact, the difficulties faced by a poorly staffed division of reform-oriented program administrators in their attempts to enforce unwelcome requirements on state and local administrators may be one reason for the fact that the reformers were able to establish the requirements. It was undoubtedly assumed by many "traditionalists" that the states and localities that so desired would simply fill out the required forms and then ignore the program requirements during implementation, thereby using the funds as general aid while complying with federal requirements on paper. Indeed, this practice was most common in the early years of Title I (Hughes and Hughes, 1972:77-80).

Had the administrative system of internal auditing in OE not been changed in 1965, it appears unlikely that Title I would have ever become tightly controlled categorical aid in the reform model except on paper in the formal requirements. It is not totally clear whether the process of tightening up will continue. However, in the case of Title I, the role of the auditors has been central to the evolution of this federal aid program.

FISCAL ACCOUNTABILITY FOR TITLE I ADMINISTRATORS: INTERNAL AUDIT AT HEW

The Health, Education and Welfare Audit Agency (HEWAA) was formed in 1965, the same year that Title I was passed by Congress. It was formed by consolidating 15 separate audit organizations that had previously been located in various program offices within HEW. However, the consolidation did not merely move the auditors in Washington into new offices, it also resulted in an expansion of the audit staffs at the 10 HEW regional offices. New auditors were hired in what was essentially a new agency in 1965. In Washington the Audit Agency was organized into five operating divisions headed by assistant directors: State and Local Audits, University and Nonprofit Audits, Social Security Installation and Management Audits, and Audit Coordination (McMickle and Elrod, 1974:111).[9] These auditors are responsible for auditing all HEW programs.

Of particular interest for this chapter is the Division of State and Local Audits, because ESEA Title I is included in this category. While HEWAA is HEW's *internal* auditor at the federal level, it functions as an *external* auditor for state and local grantees of Title I funds. The distinction is important. Internal audits are performed primarily to improve management control by providing independent and objective program information to management. External audits stress accountability (McMickle and Elrod, 1974:61-65). The GAO functions only as an external auditor. As a result, the role of the HEW auditor is even more ambiguous than is that of the GAO's personnel. While the audits of states and localities provide information to program managers, the auditors may also recommend that these grantees refund money to the U.S. Treasury if, in the auditors' opinion, the funds were misspent. Despite the fact that the HEWAA's recommendations (called "audit exceptions") are only advisory to federal program managers, they are feared by state and local administrators because of the publicity that attends them and the inconvenience of fighting their allegations. If the federal program managers (the Office of Education in this case) do not disseminate clear requirements to the state and local grantees of federal funds, or if there are different interpretations of the law between Congress and the federal administrators, one would expect that there would be significant variations in the uses to which the funds would be put at the state and local level. After the funds have been spent, the auditors investigate. The auditors who audit these states and localities must identify some standards to use in an external audit of them. As noted earlier, the GAO's *Standards* direct all government auditors to base their audits on laws, regulations, and any other relevant material. From these materials the auditors determine congressional intent, and they use that determination as the basis for audits. This procedure is followed by the HEWAA (1974).

In the first years of Title I there was little audit activity with three state audits in 1967 and 11 in 1968 (Pollen, 1975:41). No specific action was taken by OE officials on the basis of these audit reports. However, the reports were used by a group of civil rights activists who had pooled their resources to support an investigation of Title I. Using the HEWAA audit reports as the data base, a study by Ruby Martin and Phyllis McClure (1969) had a major impact on the general aid/categorical aid conflict being waged over the implementation of Title I. It did so by focusing attention upon the violations of the regulations and guidelines documented by the auditors. The Martin-McClure (1969:29) study (frequently known as the Washington Research Project) was based squarely on the reform interpretation of Title I as "change agent" that has been written into the regulations and guidelines by some administrators:

> The Title I *Regulations* are very clear about the purpose of the legislation. It is to provide educational assistance to educationally disadvantaged children in order to raise their educational attainment to levels normal for their age. Title I *programs* must be directed to the "special educational needs" of disadvantaged children. [Emphasis added]

Here is a prime example of the way in which the language used to skirt the church-state issue in 1965 affected implementation. The 1965 legislative record included discussion of programs in connection with schools although the legislative language was ambiguous on this point; this 1969 exposé emphasized the failure of many administrators to develop specific programs for targeted groups of disadvantaged students in the public schools. In other words, the auditors found that many local school systems were using Title I as general aid.

The researchers (Martin and McClure, 1969:5) documented numerous examples of general aid which fell into four basic categories:

1. Title I funds purchase services, equipment, and supplies that are made available to all schools in a district or all children in a school even though many children reached are ineligible for assistance.
2. Title I funds are spread around throughout all poverty-area schools instead of focusing on those target areas with high concentrations of low-income families.
3. Title I funds are not going to eligible children at all.
4. Title I State administration funds support non-Title I operations of State departments of education.

Such expenditures, while congruent with the desires of many congressmen, were not permissible under the regulations formulated by USOE.

The issue upon which the attention generated by the audit revelations became centered was known as "comparability." Comparability refers to the principle developed in OE that districts should expend state and local funds equally among all schools within a district before adding the Title I funds to the state and local funds in particular "target" schools for specific children. In other words, Title I money was to supplement, not supplant, state and local expenditures on eligible children. This specific issue had not been addressed in the 1965 law. Rather, the law prohibited the reduction of state aid to the districts under the following conditions:

> Sec. 207. (c) (1) No payments shall be made under this title for any fiscal year to a State which has taken into consideration payments under this title in determining the eligibility of any local educational agency in that State for State aid, or the amount of that aid, with respect to the free public education of children during that year or the preceding fiscal year.
>
> (2) No payments shall be made under this title to any local educational agency for any fiscal year unless the State educational agency finds that the combined fiscal effort (as determined in accordance with regulations of the Commissioner of that agency and the State with respect to the provision of free public education by that agency for the preceding fiscal year) was not less than such combined fiscal effort for that purpose for the fiscal year ending June 30, 1964.

These provisions of the law only concerned maintenance of the state and local fiscal effort evidenced prior to the first year's funding, and continued state expenditures for all districts whether they received Title I funds or not. Criterion 7.1 of Program Guide #44 (the most definitive of the Title I program materials) on the other hand, stated that for any given year the state and local expenditures should be generally equal among all *schools within* a district. This requirement was ignored by almost everyone until the Martin-McClure investigation focused public attention on the violations of it (Murphy, 1973:188). Congress had not originally mandated comparability; OE had formulated the provision requiring that state and local expenditures be generally comparable among all schools. (One person who had been very active on the House side asserted that no one in Congress had been concerned with strict comparability in 1965.)

School districts would not have had a problem with the "supplement, not supplant" requirement if it had not been for the "targeting" requirements, and vice versa. If school districts had been permitted to simply add federal funds on top of state and local funds in all schools (as most congressmen had intended), they would have had no difficulty. Alternatively, if they had been permitted to "target" funds to particular schools while spending less from state and local sources in those schools as they had been doing, they would have had no compliance problems. However, the combination of comparability and targeting requirements was designed to increase the concentration of funds, and this combination was tight categorical aid which resulted in the audit exceptions.

The point here is that the auditors provided the means to an end desired by some persons, and their influence could not have been predicted in 1965. Under the system of internal audits in OE before 1965, the auditors had been located in the program offices and had performed fiscal audits. This type of system was obviously what Chairman Perkins (House Hearings, 1965:195) had in mind when he said in 1965:

> I read the language in the bill just like the witness reads the language. The Department provides certain criteria to see that the money reaches the school districts we are seeking to reach, the impoverished *districts,* and at the same time, to maintain *fiscal* accounting of the funds. [Emphasis added]

Had the orientation of the HEWAA not been changing from fiscal to compliance and performance auditing, it is reasonable to assume that Representative Perkins would have gotten what he sought. However, at the same time that the reformers were establishing their interpretation of Title I in the Regulations, Guidelines, and program materials, the audits performed by the Audit Agency were expanding in scope. The early audit reports were embarrassing to the congressmen and to the members of OE who shared their general aid orientation. There was little that anyone could do, however, but seek to minimize the impact of the auditors. An OE report released in 1975 indicates that that is exactly what was done.

The "Study of the Audit Resolution Process in the U.S. Office of Education" was written by Dr. David S. Pollen, Chairman of the Title I Audit

Hearing Board. The Audit Hearing Board was established in 1972 as financial audit exceptions reported by HEWAA began piling up in OE for resolution. (One person intimately involved stated that the Hearing Board's purpose was simply to prevent the audits from becoming issues in court cases.) Pollen's study provides some interesting insights into the role of the auditors in this program.

There are several primary points to be made about the audits and the impact of the reports in OE. First, the audit exceptions are primarily based on the reform-oriented requirements that target funds within school districts, thus forcing changes at the local level if the school districts comply:

> Review of numbers of audit reports makes it evident that there are several major, recurring issues which generate a very large percentage of the Title I fiscal exceptions taken by HEWAA auditors.
>
> These issues are five in number including
>
> supplanting general aid target areas
> comparability ineligible children
>
> [Pollen, 1975:51]

Second, assuming that some reformers still remain in OE (many left during the Nixon Administration), Pollen's findings concerning support for the validity of the audit exceptions are instructive. He found that the Title I staff were much more likely to support the HEWAA's findings on noncompliance with the program requirements than were the bureau-level administrators. Pollen further noted that knowledge of the political consequences of pressing the states for repayment of disallowed funds was a major reason for high-level OE administrators' avoidance of actually requiring the funds to be repaid. They do avoid such decisions: "OE's rate of recovery of fiscal exceptions taken by HEWAA auditors is less than one percent" (Pollen, 1975:68).

Finally, Pollen's study documents a fragmented and uncoordinated "system" for audit resolution in the Office of Education (Pollen, 1975:29-37). Given the political history of Title I, it seems reasonable to conclude that such a nonsystem is the only way for OE "traditionalists" to deal with the politically sensitive audit exceptions that are based on an interpretation of Title I that both they and a number of influential congressmen find objectionable. No one can stop the auditors from auditing Title I, and it seems likely that major audit exceptions will continue to be reported. Therefore, an attempt to defuse the situation is to be expected.

CONCLUSION

In the example of ESEA Title I, HEW's internal auditors utilized the program requirements established by administrators as the primary source for their determination of the purpose of the state and local "audited entities" (congressional intent). The result in this case was to establish firmly the outlines of the program desired by the reform-oriented administrators who drafted the regulations. However, the example raises questions that extend far beyond this particular policy. Two of the most important and related questions must be faced in any discussion of administrative accountability: by whom shall administrators be held accountable, and according to what standards?

What has happened at the federal level in the United States since World War II is that federal auditors have claimed to be the primary agents of accountability, and they have chosen the standards to which they hold administrators accountable. This situation is new, and it necessarily involves the auditors in the determination of congressional intent. HEWAA turned to the regulations for this determination in Title I. However, the standards issued by the GAO raise the very real possibility of auditors' challenges to federal program administrators' interpretations of congressional intent on which those administrators have based their regulations.

Even if (as seems likely) federal auditors continue to rely heavily on regulations as the standards utilized in expanded audit coverage, their activities will become ever more important. In the case of categorical grants at least, such expanded audits will lead to the establishment of clear federal goals through the audit process as state and local administrators modify their behavior. This outcome will result because by auditing to certain criteria, the auditors will hold state and local officials accountable to those criteria for all aspects of their behavior. Although fiscal exceptions are rarely upheld, the adverse publicity and anxiety faced by a noncompliant grant recipient will tend over time to result in program modifications designed to satisfy the auditors. As such situations develop, since the auditors do not have time to audit all aspects of all programs, their choice of criteria and their operationalization of those criteria will define the programs for grant recipients. Effective fiscal accountability will be defined by the auditors, and the system of interest group liberalism will feel the effects of the new input.

Federal auditors now have the statutory authority to investigate virtually every aspect of administrative behavior that involves federal funds. State audit agencies are acquiring more of the same powers over state funds. "Accountability" is a slogan utilized by the auditors to justify this expansion of their activities. For the auditors, however, accountability has a very definite meaning: it refers to lower-level administrative behaviors judged by the auditors to be in compliance with higher-level administrative directives issued to implement legislative intent. It is a hierarchical, management-oriented view of accountability that may be at variance with the desires of analysts who seek decentralized, client-oriented accountability on the part of administrators. The word is the same; the implementation of it is very different.

The role of the auditors must be considered when dealing with accountability because their importance, while already great, is growing. Most program administrators are not afraid of community groups; they are afraid of the federal auditors. Auditors can impose meaningful sanctions: internal audit agencies recommend and the GAO can require that federal funds be returned to the United States Treasury. The auditors are prime examples of the "feds" of whom state and local officials speak with distaste. Currently they are functioning as the primary agents of accountability. Their activities should be acknowledged. Fiscal accountability has been broadly defined; the new situation has not been widely recognized.

NOTES

1. The auditors (for example, Staats, 1975:64) consider only the traditional audits as focusing on *fiscal* accountability, with management audits addressing *managerial* accountability, and performance audits focusing on *program* accountability. However, the government's auditors do all of the audits, and their legitimacy is based upon their link to expenditures. Therefore, for purposes of this paper, fiscal accountability refers to the goal of the auditors' investigations, whether broadly or narrowly structured.

2. Both Elmer Staats and E. L. Normanton are public auditors. Staats is the Comptroller General of the General Accounting Office and Normanton is a member of the British Exchequer and Audit Department.

3. The GAO's Standards do not explicitly state that the auditors should make a determination of congressional intent. However, in the section dealing with "Legal and Regulatory Requirements" (Comptroller General, 1972:28-31), it is noted that the auditors must determine the "purpose of the entity" audited. In various discussions with auditors, it became clear that they think in terms of "congressional intent."

4. This case study had its origin in a study of the passage and implementation of ESEA Title I (Stoner, 1976). The information presented is based on that research, during the course of which the importance of the auditors became evident. The data utilized was of two types: (1) public documents and published records, and (2) 89 open-ended interviews conducted with persons connected with Title I at the federal, state, and local levels.

5. See Munger and Fenno (1961) for a history of previous attempts to pass general aid to education.

6. Price (1961) contains a careful account of the manner in which the religious issue affected the Kennedy aid to education bill.

7. This approach was justified by many on the basis of *Everson* v. *Board of Education.* In a 5-4 vote the Supreme Court upheld the constitutionality of a New Jersey state law that subsidized the costs of transportation of children to nonpublic schools. The Court ruled that the primary object of the law was public safety, not aid to private education, and that therefore it did not violate the First Amendment.

8. See Stoner (1976:35-80) for a detailed analysis of these hearings. Senator Robert Kennedy was not an enthusiastic supporter of Title I because he did not share the faith

of his colleagues that more money for education would have much impact on students from poor homes. However, most of his influence was exerted before the bill reached Congress (McLaughlin, 1973:3-7).

9. The Health, Education and Welfare Audit Agency was placed under the direction of an Inspector General in 1977. It is not yet clear what impact this change in organizational structure in Washington will have on the performance of audits in the regions or the resolution of audits in Washington.

REFERENCES

BAILEY, S. K., and MOSHER, E. K. (1968). ESEA: The Office of Education administers a law. Syracuse, N.Y.: Syracuse University Press.

DITTENHOFER, M. A. (1970). "Is auditing a fourth power of government?" State Government, 43(Summer):179-183.

EDELMAN, M. (1964). The symbolic uses of politics. Urbana: University of Illinois Press.

EIDENBERG, E., and MOREY, R. (1969). An act of Congress. New York: W. W. Norton.

FITZGERALD, M. J. (1975). "The expanded role of the General Accounting Office." The Bureaucrat, 3(January):383-400.

GRAHAM, R. V. (1970). "Is auditing a fourth power? Yes." State Government, 43(Autumn):258ff.

HUGHES, J. F., and HUGHES, A. O. (1972). Equal education. Bloomington: Indiana University Press.

KNIGHTON, L. M. (1970). "Is auditing a fourth power? No." State Government, 43(Autumn):258-265.

LEE, G. C. (1949). The struggle for federal aid. New York: Teachers College, Columbia University.

LOWI, T. J. (1969). The end of liberalism. New York: W. W. Norton.

MARTIN, R., and McCLURE, P. (1969). Title I of ESEA: Is it helping poor children? Southern Center for Studies in Public Policy and the NAACP Legal Defense and Education Fund (August).

McLAUGHLIN, M. W. (1973). "Evaluation and reform: The case of ESEA Title I." Ph.D. dissertation, Harvard University.

McMICKLE, P. L., and ELROD, G. (1974). Auditing public education. Montgomery, Ala.: Alabama Department of Education.

MUNGER, F. J., and FENNO, R. F. (1961). National politics and federal aid to education. Syracuse: Syracuse University Press.

MURPHY, J. T. (1973). "The education bureaucracies implement novel policy: The politics of Title I of ESEA, 1965-72." Pp. 160-198 in A. P. Sindler (ed.), Policy and politics in America. Boston: Little, Brown.

NORMANTON, E. L. (1966). The accountability and audit of governments. New York: Frederick A. Praeger.

POIS, J. (1975). "Trends in General Accounting Office audits." Pp. 245-277 in B. L. R. Smith (ed.), The new political economy. New York: John Wiley.

POLLEN, D. S. (1975). "Study of the audit resolution process in the U.S. Office of Education." Unpublished internal USOE document.

PRICE, H. D. (1961). "Race, religion and the Rules Committee," in A. F. Westin (ed.), The uses of power. New York: Harcourt, Brace & World.

SHARKANSKY, I. (1975). "The politics of auditing." Pp. 278-318 in B. L. R. Smith

(ed.), The new political economy. New York: John Wiley.
STAATS, E. B. (1975). "New problems of accountability for federal programs." Pp. 46-67 in B. L. R. Smith (ed.), The new political economy. New York: John Wiley.
STONER, F. E. (1976). "The implementation of ambiguous legislative language: Title I of the Elementary and Secondary Education Act." Ph.D. dissertation, University of Wisconsin-Madison.
THOMAS, N. C. (1975). Education in national politics. New York: David McKay.
U.S. Code Congressional and Administrative News 29 (1965). Public Law 89-10(March 26, 1965).
U.S. Comptroller General (1976). "An overview of federal internal audit" (November).
--- (1972). "Standards for audit of governmental organizations, programs, activities and functions."
U.S. Congress, House (1965). Aid to elementary and secondary education, Hearings before General Subcommittee on Education on H. R. 2361 and H.R. 2362. 89th Congress, 1st session.
---, Senate (1965). Elementary and Secondary Education Act of 1965, Hearings before Subcommittee on Education on S370. 89th Congress, 1st Session.
U.S. Congressional Record (1974). 120:6267. 93d Congress, 2d session.
--- (1965). 111:7317. 89th Congress, 1st session.
U.S. Department of Health, Education and Welfare (HEW), Audit Agency (1974). HEW Audit Agency Policy Handbook. Washington, D.C. (August).
U.S. General Accounting Office [GAO] (1974). "Internal auditing in federal agencies: Basic principles, standards, and concepts" (August).

Part IV

Research Designs for Studying Accountability

Introduction

☐ THE NEXT TWO CHAPTERS go together. In the first David Nachmias states and analyzes the classical research design for testing propositions. He then indicates variations and, essentially, ways in which the rigorous assumptions may be relaxed in order to bring the method within the reach of most studies of accountability. It is notable that accountability and evaluation studies are, in a sense, the same. If we go to the most global level, we must ask: Is the program indeed effective in achieving the goals its sponsors wished? At what cost compared to the null case, where no program was applied, and to what profit?

Thomas E. James and Ronald D. Hedlund have written an article which, in part, translates the Nachmias argument into the realities of practice in ordinary terms–daily life in the program. It is, in a sense, a useful checklist one should keep in mind when facing the whole promising, complex, and often frustrating demand for accountability through evaluation of results.

Robert K. Yin argues that the difficulties in applying experimental design are so great as to vitiate it. He suggests the need to develop other logics of proof and disproof, and other methods for using such logics. Not fancifully, he suggests we look at the ways in which truth values are assigned in areas as diverse as the study of checks and balances in political science–translatable into domestic governmental architecture on one hand, international relations on the other. One will evaluate such a criticism and such alternatives in the light of his own received methodological gospel, but the point is one that deserves pondering.

12

Assessing Program Accountability: Research Designs

DAVID NACHMIAS

□ THE CONCEPT OF PUBLIC ACCOUNTABILITY is essentially the concept of democracy. It embodies the idea that government should be responsible to the people. The discharge of that responsibility requires that policy makers have credible information about the operation and effectiveness of programs brought into being by federal, state, or local legislators in responses, presumably, to the wishes of their constituencies.

Traditionally the kinds of information collected for assessing public accountability were restricted to fiscal accountability and management accountability. For example, the General Accounting Office has defined its roles as focusing primarily on the quality of accounting practices maintained by the operating agencies (fiscal accountability), and the efficiency with which resources are used (management accountability). Notwithstanding the significance of such concerns, they tell little, if anything, of the extent to which public programs accomplish their objectives. The realization that program effectiveness is a distinct and significant aspect of accountability led to the adoption of the Legislative Reorganization Act of 1970. In its explanation of Section 204(a) of the Act, the House Committee on Rules stated that the intent was that the Comptroller General "shall review and analyze program results in a manner which will assist Congress to determine whether those programs and activities are achieving the objectives of the law" (House Committee and Rules, 1970:82).

The broadening concept of public accountability to include program effectiveness is by no means limited to the GAO. In the early 1970s, about

200 new program-effectiveness investigations began each year with direct federal support and with average budgets of $100,000 each (Freeman, 1977). This, however, has not resulted in the generation of systematic and verified information upon which policy makers could lean. Among the reasons that policy makers do not utilize program-effectiveness information, Horst et al. (1976) point out the following methodological ones: (1) different studies of the same program are not comparable; (2) program-effectiveness studies have failed to provide an accumulating, accurate body of evidence; and (3) program-effectiveness studies often address unanswerable questions and produce inconclusive results.

These shortcomings stem from the reluctance to explicitly accept the argument that at the heart of all program-accountability research activities is the idea of causality. That is, a public program is expected to produce a change in the target population in the direction and magnitude intended by the policy makers.

In basic social science research the tendency to shy away from causal investigations has deep philosophical roots (Francis, 1961). Perhaps the major problem with establishing causal relations is that causality belongs to the theoretical-abstract domain, and consequently cause-and-effect relationships cannot be empirically observed: "One thinks in terms of a theoretical language that contains notions such as causes, forces, systems, and properties. But one's tests are made in terms of covariation, operations and pointer reading" (Blalock, 1964:5). In applied social research including program-effectiveness investigations, the gap between the theoretical and the empirical language is less wide and can more easily be bridged with simplified impact models that are subject only to few limitations and that apply to the real world. The researcher can start with a finite number of specified program and target variables and construct an explicit impact model upon which both action and research can be based (Nachmias, 1978). Having constructed such an impact model, one proceeds in two directions: formulation of predictions that can be translated into testable hypotheses, and simultaneously setting operational procedures that can be applied to test the hypotheses. Such operational procedures are embodied in the notion of research designs.

Research designs are operational models of proof for inferring cause-and-effect relations; as such they are central to the conduct of program accountability research. In the following section the logical structure of research designs is discussed and exemplified. Next, the focus is on the components of research designs and on sources of validity. Policy experiments, quasi-experimental designs, and structural equation models are discussed in later sections with the intent to emphasize their advantages and inherent limitations for program accountability research.

THE LOGIC OF RESEARCH DESIGNS

The classic research design for the study of program accountability consists of two groups: an experimental group and comparison group. These two groups are equivalent, except that the experimental group is introduced to a program and the comparison group is not. To evaluate the effectiveness of a program, measurements on the target variable(s), designated as scores, are taken twice from each group. One measurement, the pre-program, is taken prior to the implementation of the program in the experimental group; a second measurement, the post-program, is taken after the program has taken place. The difference in scores between the post-program and pre-program is compared in each of the two groups. If the difference in the experimental group is significantly larger than in the comparison group, one can infer that the program had caused the change in the target variable(s). This design can be diagramed as in Table 1 where X symbolizes the program; O_i measurements on the target variable, and d_e and d_c stand for the differences between the post-program and pre-program measures.

One interesting application of this design is the Manhattan Bail Project initiated by the Vera Institute in New York (Boteim, 1965). The chief objective of the project was to furnish criminal court judges with evidence regarding the contention that many persons could be successfully released prior to trial and without bail if verified information concerning defendants' backgrounds were available to the court at the time of bail determination. To assess the accountability of the program, a target group including persons accused of felonies as well as misdemeanors was selected; individuals charged with very serious crimes were excluded from the study. New York University law students and Vera staff reviewed the defendants' records of employment, family, residences, references, current charges and previous records in order to decide whether a pre-trial release without bail should be recommended to the court. The total group of recommendees was split randomly into experimental and comparison groups, and recommendations were made to the judge in the experimental group only. The target variables were pre-trial release granted, case dispositions, sentences and default rate. Judges in the first year granted parole to 59% of the recommended defendants, compared to only 16% in the control group; recommendations based on information then

TABLE 1
THE CLASSIC EXPERIMENTAL DESIGN

	Preprogram		Postprogram	Difference
Experimental Group	0_1	X	0_2	$0_2 - 0_1 = d_e$
Comparison Group	0_3		0_4	$0_4 - 0_3 = d_c$

served to increase the rate of release without bail. Of the recommended group, 60% was either acquitted or had their cases dismissed, compared to 23% of the control group. During 1961-1964, less than 1% of the experimental group failed to show up in court for trial–a rate that was considerably lower than that for similarly charged defendants who had posted bail–suggesting that the relaxation of the bail requirement did not result in unacceptable default rates. Based on these findings, the accountability of the program was established, and the New York Probation Department extended it to criminal courts in all five boroughs of the city.

ELEMENTS OF RESEARCH DESIGNS

The classic research design consists of four essential elements: comparison, manipulation, control, and generalizability.

COMPARISON

The process of comparison underlies the concept of covariation or association between a program and its target variable(s). Suppose a relationship exists between a program and the rates of unemployment; one would then expect to find a joint occurrence of both the program variables and certain rates of unemployment. That is, communities will have lower unemployment rates after implementing the program than before. Similarly, communities that implemented the program will have lower unemployment rates than communities that did not implement it. Thus, to assess the joint occurrence of a program and changes in the target variable(s), a comparison is made of a group that was introduced to the program with one that was not, or of the group's scores on the target variable(s) before and after the implementation of the program. In the former case, an experimental group is compared with a comparison group; in the latter, a group is compared with itself.

MANIPULATION

The notion of program accountability implies that a program has changed, and usually has improved, certain conditions. A program such as Head Start is accountable if it improves the cognitive skills of children participating in it. In other words, if a program is actually effective, then individuals or other units (e.g., groups, communities, cities) introduced to it should change over the time of participation. The assumption that in the real world there are certain agents that produce change has its counterpart in the laboratory experiment in which the researcher actually acts as such an agent. The notion is "that if X is a cause of Y, and if it were possible to hold constant all other causes of Y, an experimental manipulation of X (i.e., an externally produced change in X) should be accompanied by an observed change in Y" (Blalock, 1964:22). The change in Y (the target variable) may not occur before the change in X (the program) if one is to infer that Y is caused by X. Indeed in program accountability research, experimental manipulation should concern only the target variable(s) and not some other unsuspected variables.

CONTROL

The third essential element of a research design is that other factors be ruled out as rival explanations of the observed relationship between a program and its target variable(s). Such factors might invalidate inferences concerning the effectiveness of a program. Campbell and Stanley (1966) have formulated this as the problem of internal validity. It refers to the degree to which a research design allows one to rule out alternative explanations for the way in which a particular program variable is causally related to target variable(s). The factors that may jeopardize internal validity can be classified into those which are extrinsic to the research operation and those which are intrinsic and impinge upon the results during the study period.

Extrinsic factors refer to the possible biases resulting from the differential recruitment of individuals or other units to the experimental and comparison groups. As an illustration, consider Chapin's (1940) early study on the social effects of public housing. This investigation was an attempt to examine the changes occurring in the social life of slum families as a result of their rehousing in public housing projects. Chapin compared an experimental group of families who had been rehoused with a comparison group of families who were still living under slum conditions. The main findings of the study showed a marked improvement in the social life of the experimental group, leading to the inference that public housing projects are effective programs because they change the life style of their inhabitants. However, one rival explanation of the observed change in the rehoused families is that the people in new housing projects were initially different from the families serving as a comparison group. Perhaps the groups differed in type of employment, level of education, size of family, or social orientations. These factors could have accounted for the observed differences between the two groups. Selection factors such as these must be controlled before the investigator can rule them out as rival explanations.

Intrinsic factors refer to changes in the studied persons, groups, or other units of analysis during the investigation period, changes in the measuring instruments, or the reactive effects of the study itself. The following are the major instrinsic factors that might invalidate inferences concerning the effectiveness of a program (Campbell and Stanley, 1966; Campbell, 1969).

History. This factor refers to all events occurring during the time of the investigation that might effect the studied individuals and provide rival explanations for the change in the target variable(s). The longer the time lapse between the pre-program and the post-program measurements, the higher the probability that events other than the program variable(s) might become potential rival explanations.

Maturation. A second group of factors that may become plausible rival hypotheses is referred to as maturation and includes processes that produce change in the target variable merely as a function of the passage of time. Suppose one wants to evaluate the effect of a specific teaching program on student achievement and records achievement before and after the implementation of the program. Between the pre-program and the post-program measurements, students have gotten older; this change, unrelated to the teaching program, could possibly explain the difference between the two tests. Processes such as growth and fatigue can produce changes in the target variable independently of the program.

Testing. The possible reactivity of measurement is a major problem in program accountability research. The process of testing (or measuring) may change the phenomena being tested. The effect of being pre-tested might sensitize individuals and improve their scoring on post-program measures. A difference between post-program and pre-program scores could thus be attributed not necessarily to the program but rather to the experience gained by individuals while participating in the pre-program encounter. For example, an ineffective program to improve cognitive skills could report improvement if identical tests were used in the post-program and pre-program measurements, since individuals might remember items or questions and discuss them. One might question whether a program would have had the same effect if the pre-program measurements had not sensitized individuals to the program objectives.

Instrumentation. Instrumentation or instrument decay designates changes in the measuring procedures and/or instruments between the pre-program and the post-program. To associate the difference between post-program and pre-program scores with a program, one needs to assume that repeated measurements with the same instrument under unchanging conditions will yield the same results. If such an assumption cannot be made, observed differences could be attributed to the change in the measurement instrument and not necessarily to the program. For example, if a program to improve cognitive skills were evaluated by comparing pre- and post-program ratings by psychologists, any changes in their standard of judgment which occurred between testing periods would bias the findings.

Regression Artifacts. These are pseudo-shifts occurring when individuals have been selected upon the basis of their extreme scores on pre-program measures. Regression effects "guarantee that individuals who scored below average on a pre-test will appear to have improved upon retesting" (Houston,

1972:60). Relatedly, if the measuring instrument is not perfectly reliable, units who scored above the average on the pre-program would appear to have done less well upon remeasuring.

Selection. This refers to biases that might result because of the procedure by which individuals or other units of analysis were assigned to the experimental and comparison groups. When the selection procedures are improper, post-program differences between the experimental and comparison groups reflect initial differences rather than program effectiveness. Selection effects are especially salient and problematic if the individuals themselves determine whether to participate in a program. In such cases, one cannot tell whether it is the program itself which caused the observed differences between the experimental and comparison groups, or whether the observed effects.

Experimental Mortality. Experimental mortality refers to the differential loss of persons from the experimental and comparison groups. Such differential loss may create observed differences in the post-program measures; these, however, cannot be attributed to the program. For example, in a study on the effect of the media on prejudice, if most dropouts were prejudiced individuals, the impression rendered could be that exposure to media reduced prejudice, whereas in fact, it was the effect of experimental mortality that produced the observed shift in opinion.

Selection-Maturation Interaction. This refers to the selection of individuals whose characteristics will change between the pre-program and the post-program regardless of whether they participate in a program.

RANDOMIZATION

Experimental designs, such as the classic one, provide control against extrinsic and intrinsic sources of internal invalidity. Ideally, the comparison and experimental groups are under identical conditions during the study, except for their differential exposure to the program; persons in the comparison group do not participate in it. Consequently, features of the experimental situation or external events that occur during the experiment are likely to influence the two groups equally and will not be confounded with the effects of the program. For example, history cannot remain a rival hypothesis since the experimental and comparison groups are both exposed to the same events occurring during thc investigation. Similarly, maturation is neutralized since the two groups undergo the same changes. Although the inclusion of a comparison group does not necessarily avoid the mortality problem, since the loss of cases might be differential and bias the results, an acceptable procedure is to include in the final analysis only cases for which complete information is available. Thus, the ideal experiment is one in which all the factors likely to affect the program outcomes are controlled.

In practice, approximations to the ideal experiment are attempted through the method of randomization. Randomization refers to the assignment of individuals or other units of a target population to groups in such a way that for any given assignment to a group, every member of the target population has an equal, nonzero probability of being selected for that assignment. In other words, the underlying rationale of randomization is that since in random procedures every member of a target population has an equal, nonzero probability of being chosen, members with certain distinct attributes (lower or upper social class, white or black, urban or rural areas, and so on) will, if selected, be counterbalanced in the long run by the selection of other members of the target population with the "opposite" quality or quantity of the characteristic.

Randomization makes a program accountability study unbiased, since it can have no a priori tendency to favor one group of experimental individuals over another. In large program experiments randomization tends to equate groups with respect to all possible antecedents factors. Also, randomization permits a valid estimate of variability due to error which is needed to specify the stability of inference (Cochran and Cox, 1957). Third, randomization, when executed with a representative sample, enables one to extend his or her inferences beyond the data at hand. This last point leads to the fourth essential element of a research design, namely, generalizability.

GENERALIZABILITY

Internal validity is a crucial aspect of program accountability research. An additional significant issue concerns the extent to which one's findings can be generalized to larger populations and be applied to other settings. This problem is termed the external validity of research designs. Two sources of external validity concern the representativeness of the sample and the reactive arrangements in the research procedure.

Representativeness of the Sample. The random assignment of individuals or other units to experimental and comparison groups assures equality between the groups and thus contributes to the internal validity of the study. However, it does not necessarily assure representativeness of the target population. Results that prove to be internally valid might be specific to the sample selected for the particular study. To enable the making of generalizations beyond the limited scope of the specific study, care should be taken to select the sample using a sampling method that assures representation. Probability methods such as random sampling would make generalizations to larger and clearly defined populations possible.

Reactive Arrangements. The results of a study are to be generalized not only to a larger population but also to a real-life setting. This cannot always be accomplished, especially when a study is carried out in a highly artificial and contrived situation. In addition to the possible artificiality of the experimental setting, various features in the setting might be reactive and affect the external validity of the study. For example, the pre-test may influence the responsiveness of individuals to the experimental stimulus; its observed effect would thus be specific to a population that has been pre-tested.

Research designs for assessing program accountability can be classified by the extent to which they meet the validity criteria discussed so far. Some designs allow for program manipulation but fail to employ methods of control or an adequate sampling frame; others may include one or more control groups but have no control over the implementation of the program. Accordingly, it is possible to make a distinction between three major types of research designs: experimental, quasi-experimental and pre-experimental. In experimental designs, individuals or other units of analysis are randomly assigned to the experimental and control groups. Such designs allow for comparison, control, manipulation, and, in most cases, generalizability. A quasi-experimental design is one in which one or more extrinsic and/or intrinsic factors are not being controlled for. Typically, this occurs when randomization cannot be achieved. To compensate for this, supplementary data and specialized techniques of data analysis are used. Pre-experimental designs include even fewer safeguards and inferences concerning program effectiveness depend primarily on the quality and the inclusiveness of the data analyses performed. The following sections elaborate on the logic and the mechanics of these three general types of research designs.

CONTROLLED EXPERIMENTATION

The classic experimental design presented earlier is one of the strongest models for assessing program effectiveness. It allows for pre- and post-program measurements, and comparison-group–experimental-group comparisons; it permits the manipulation of a program and, by including randomized groups, it controls for most sources of internal validity. However, the classic experimental design is relatively weak on external validity. There are two variations of the classic design that are stronger in this respect: the Solomon Four-Group design and the Post-program-Only Comparison Group design.

THE SOLOMON FOUR-GROUP DESIGN

The pre-program measurements in an experimental setting have advantages as well as certain limitations. Although they provide an assessment of the time sequence as well as a basis of comparison, they can have severe reactive effects. Pre-program sensitization introduces the possibility that the application of pre-program measurement in and of itself might effect the findings.

TABLE 2
THE SOLOMON FOUR-GROUP DESIGN

	Preprogram		Postprogram
R	O_1	X	O_2
R	O_3		O_4
R		X	O_5
R			O_6

For example, measuring public attitudes toward a government program prior to the introduction of the program may sensitize individuals to respond in a way different than non-pre-tested persons would respond. Furthermore, there are circumstances under which it might be impractical to administer pre-program measures.

The Solomon Four-Group design presented in Table 2, contains the same components as the classic design, plus an additional set of comparison and experimental groups that are not pre-measured (Solomon, 1949). Therefore, the reactive effect of measurement can be directly assessed by comparing the two experimental groups ($O_2 - O_5$) and the two comparison groups ($O_4 - O_6$). The comparisons will indicate whether X (the program) has an independent effect on the groups that were not sensitized by pre-program measurement. If it can be shown that the program had an effect even with the absence of pre-program measures, the results can be generalized to target populations that were not measured prior to the implementation of the program. Moreover, as Campbell and Stanley (1966:25) suggest, "not only is generalizability increased, but in addition, the effect of X is replicated in four different fashions: $O_2 > O_1$, $O_2 > O_4$, $O_5 > O_6$ and $O_5 > O_3$ . . . if these comparisons are in agreement, the strength of the inference is greatly increased."

Lana (1969) has demonstrated through a series of experiments that across a wide variety of attitudes and opinions, either (1) there was no difference in experimental effects between pre-tested and post-tested groups, or (2) where differences were found, it was shown that smaller changes occurred for the pre-tested than for the post-tested groups. That is, if anything, pre-tests tended to result in under- rather than overestimates of effects. These findings led to the conclusion that while pre-program sensitization may logically threaten the internal and external validity of program experiments, the actual effects are rather small (Bernstein, 1976:116). Accordingly, a decision whether to employ the Solomon Four-Group design for program accountability research when the concern is with generalizing to wider populations must depend on the merits of each and every case. For certain programs, the population of interest will itself be a pre-tested one. For example, participants in some programs are required to fill out forms and questionnaires prior to participation, in much the same way as is required of individuals participating in laboratory experiments. In such cases the use of the Solomon Four-Group design would enhance the external validity of the causal inferences.

POST-PROGRAM-ONLY COMPARISON GROUP DESIGN

The Solomon Four-Group design is a strong experimental design. However, using such an elaborate design is often impractical, too costly, or the pre-program measures might be reactive. The Post-Program-Only Comparison Group design is a variation of both the classic design and the Solomon design; it omits the pre-measured groups altogether. The design is illustrated in Table 3.

The Post-Program-Only Comparison Group design is identical to the two last groups of the Solomon Four-Group design, which are not premeasured. Individuals are randomly assigned to either the experimental or the comparison group and are measured during or after the implementation of the program. As an illustration, suppose a researcher examining the effects of a racist film on racial prejudice selects a sample that is randomly assigned to the two groups. One group is shown the film and later the two groups are interviewed. To assess the effect of the film on racial prejudice, responses to the interview in the two groups are compared. A significant difference will indicate that the film had an effect on prejudice. This is evaluated by comparing the occurrence of prejudice in the experimental group with its occurrence in the comparison group. The time-order can be inferred from the randomization process used to assign the individuals or other units to the different groups. This procedure, as previously suggested, removes any initial differences between the groups; therefore, it can be inferred that the observed difference was produced by the program.

TABLE 3
THE POSTPROGRAM-ONLY CONTROL GROUP DESIGN

		Postprogram
R	X	0_1
R		0_2

The Post-Program-Only Comparison Group design controls for most intrinsic sources of validity. With the omission of pre-program measurement, testing and instrumentation are no longer relevant sources. It can also be assumed that the remaining intrinsic factors are controlled, since both groups are exposed to the same external events and undergo the same maturation processes. In addition, the extrinsic factor of selection is controlled by the random assignment of individuals, which removes an initial bias of either group.

FACTORIAL DESIGNS

The term factorial design refers to controlled program experiments that involve several program variables and in which the variables (or factors) are orthogonal to one another. Orthogonal means that each of the levels of each factor is manipulated or administered in combination with each of the levels of the others. Factorial designs elaborate on the number of experimental and comparison groups in view of the fact that, as Fisher (1937:101) has suggested:

> We are usually ignorant which, out of the innumerable possible factors, may prove ultimately to be the most important, though we may have strong presuppositions that some few of them are particularly worthy of study. We have equally no knowledge that any one factor will exert its effects independently of all others, . . . or that its effects are particularly simply related to variations in these other factors.

The chief advantage of factorial designs is that they may considerably broaden the range of generalizability. Instead of "controlling for everything," as in the case of single-variable designs, additional relevant variables are being introduced, each at two or more different levels. Consequently, one is not restricted by some constant levels of each of these relevant variables when generalizing on the effect of a program. Rather, one is in a position to infer that the effect of a program occurs similarly across several levels of the other variables or, alternatively, that the effect is different at different levels of one or another of these variables.

In factorial designs the effect of one of the program variables at a single level of another of the other variables is referred to as "simple effect." The over-all effect of a program variable, averaged across all of the levels of the remaining variables, is referred to as its "main effect." Interaction describes the manner in which the simple effects of a variable may differ from level to level of other variables. For example, if the simple effects of a variable A differ significantly from level to level of another variable B, this is expressed by demonstrating that the interaction between A and B was significant.

To exemplify the logic of factorial designs, consider a program involving

TABLE 4
A 2^2 FACTORIAL DESIGN

	a_1	a_2
b_1	a_1b_1	a_2b_1
b_2	a_1b_2	a_2b_2

two orthogonal factors A and B, each with two levels—a_1, a_2 and b_1, b_2—this is a 2^2 design. Table 4 illustrates this design and the resulting four combinations.

Suppose two communities have been assigned at random to each of the four cells. This means that each community will get a combination of two experimental manipulations, but each pair of communities will get a different combination. Now, it is possible to conceive the factors as being independent; that is, two separate experiments are actually being administered with the same sample. In one experiment, program A is being introduced and in the other, program B. The main effect of B is contained in the difference between the mean (X) of cells a_1b_1 and a_2b_1 combined versus cells a_1b_2 and a_2b_2 combined. Similarly, the main effect of A is contained in the difference between the mean of cells a_1b_1 and a_1b_2 combined versus cells a_2b_1 and a_2b_2 combined. The simple effect of B with a_1 involves the difference between the a_1b_1 and a_1b_2 cell means, whereas the simple effect of B with a_2 involves the difference between the a_2b_1 and a_2b_2 cell means. The difference between these two differences defines the B x A interaction:

$$\left[A_{a_1b_1} - X_{a_1b_2}\right] - \left[X_{a_2b_1} - X_{a_2b_2}\right]$$

QUASI-EXPERIMENTAL DESIGNS

The controlled experiment allows the most unequivocal assessment of the extent to which a program is accountable. However, social, political, and ethical considerations may impede on the application of controlled experimentation. In recent years considerable attention has focused on developing quasi-experimental research designs where true experimentation is either impossible or impractical. A quasi-experimental design is one in which some potentially significant confounding factors are not being controlled for. To compensate for this weakness, supplementary data and specialized techniques of data analysis are being used to reduce ambiguities of inference. The present section centers on the weakest and the strongest quasi-experimental designs.

CONTRASTED-GROUPS DESIGNS

A cardinal problem in program accountability research develops when opportunities for randomization are unavailable. At times, intact comparison groups are being employed either at the pre-program period only or at the post-program period. Indeed, inferences concerning the causal effects of a program are especially vulnerable when the compared groups are known to differ in some important attributes. For example, when comparing poor communities with relatively well-to-do ones, or when comparing groups from different racial and/or ethnic backgrounds. In situations where differences among such contrasted groups have to be assessed, several elaborations in the research design are possible. These serve as safeguards against the intrusion of confounding factors.

The least elaborated type of design for contrasted groups is that in which the only "treatment" consists of being a member in categoric groups; members of each group are measured with respect to the target variable. An example would be comparing the reading performance of children residing in, say n, different neighborhoods. Differences in scores obtained for the above n groups are amenable to straightforward comparative statistical analyses (e.g., difference between means). However, because such contrasted groups differ from one another, difficulties arise when attempting to assess the causes for the observed differences. Relatedly, the groups might differ from one another because of artifacts in the measurement procedures and not because of real differences (Philips, 1971). One way to reduce the risk of being wrong when making inferences based on Contrasted-Groups designs is to obtain supplementary evidence over time regarding the hypothesized differences. Thus, if the same finding is obtained in other places and comparisons are made on a wide variety of measures, then such supplementary evidence will increase the inferential powers of the Contrasted-Group design.

In some cases measures are available on a number of occasions before and after the implementation of the program. In such situations multiple measures before and/or after the implementation of the program can be obtained. The chief advantage of multiple measures is that they provide an indication of the amount of normal variation in the target variable from time to time, irrespective of the program's impact. Evidence for program effectiveness consists of a sharp interaction from before to after the implementation of the program for the groups being compared, as is illustrated in Figure 1.

Unlike the hypothetical results in Figure 1, the illustrative findings in Figure 2 suggest that the program had no effect at all on the individuals in group A beyond what could be expected from the usual course of events, as is evidenced in group B. The apparent change in group A is illusory because it is matched by a proportional change in group B. Unless striking results are obtained like those depicted in the figures, usually it is risky to make any

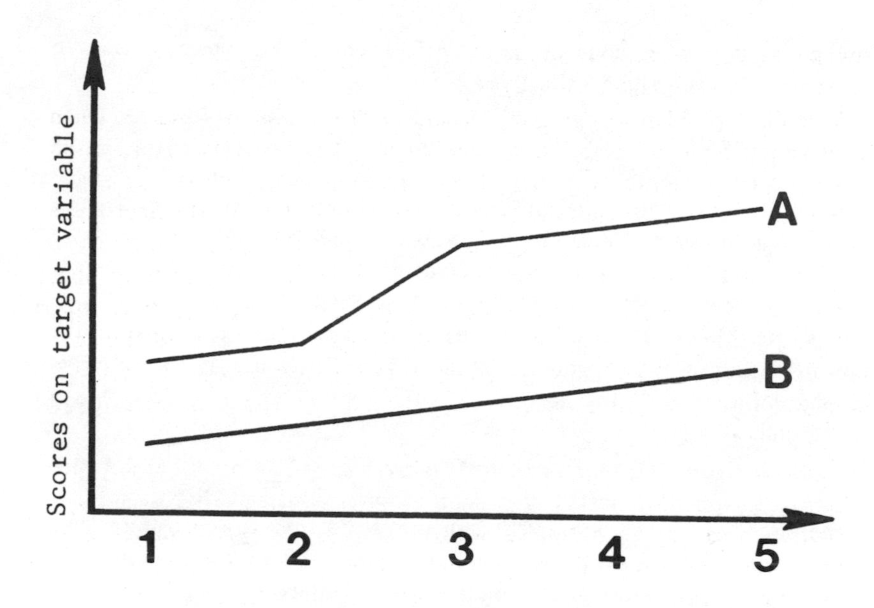

FIGURE 1
COMPARISON OF TWO CONTRASTED GROUPS INDICATING THAT THE PROGRAM HAD AFFECTED THE TARGET VARIABLE

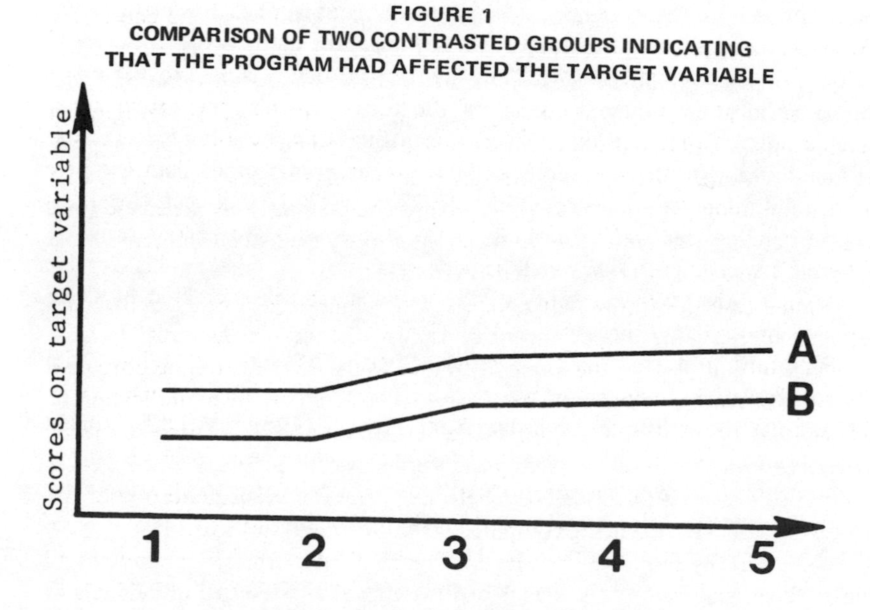

FIGURE 2
COMPARISON OF TWO CONTRASTED GROUPS WHICH INDICATES THAT A PROGRAM HAD NO EFFECT

firm causal inferences concerning the effectiveness of a program from a Contrasted-Groups design (Nunnally, 1975).

TIME-SERIES DESIGNS

In some cases, no comparison group at all is available for comparing the effects of a program. Research designs where pre- and post-measures are available on a number of occasions before and after the implementation of a program are commonly referred to as Time-Series designs. Usually one attempts to obtain at least three sets of measures before and after the implementation of the program. A typical Time-Series design can be represented as follows:

$$O_1 \ O_2 \ O_3 \ X \ O_4 \ O_5 \ O_6$$

The employment of such a design makes it possible to separate reactive measurement effects from the effects of a program. A Time-Series design also enables one to see whether a program has an effect over and above the reactive effects. The reactive effect shows itself at O_3; this can be contrasted with O_4. If there is an increase at O_4 over and above the increase at O_3, it can be attributed to the program. A similar argument applies for maturation. However, Campbell and Stanley (1966:36) maintain that history is the most serious problem with this design. Their argument is that it is plausible that the program did not produce a change in the target variable, but rather some other event or combinations of events occurring during the implementation period. Indeed, if there are constantly recurring events other than the program, valid inferences concerning its effectiveness are risky at best. Nevertheless, in concrete research situations such extraneous factors might show up between, say, O_2 and O_3, as well as between O_3 and O_4, making history less of a threat. Also, the credibility of the inference can be increased by using supplementary data if these are available.

One serious problem with Time-Series designs is the fact that all time-series are unstable even when no programs are being implemented. The degree of this normal instability is, according to Campbell (1969:413) "the crucial issue, and one of the main advantages of the extended time-series is that it samples this instability." Regression artifacts also present a serious threat to the validity of Time-Series designs, especially when these are characterized by instabilities. With any highly variable time-series, if one is to select a point that is the "highest so far," the next point on the average will be lower, or nearer the general trend.

Figure 3 is illustrative of a case where it can be concluded that a program had no effect on the target variable. The curve goes up from before the introduction of a program to after its implementation. However, the curve

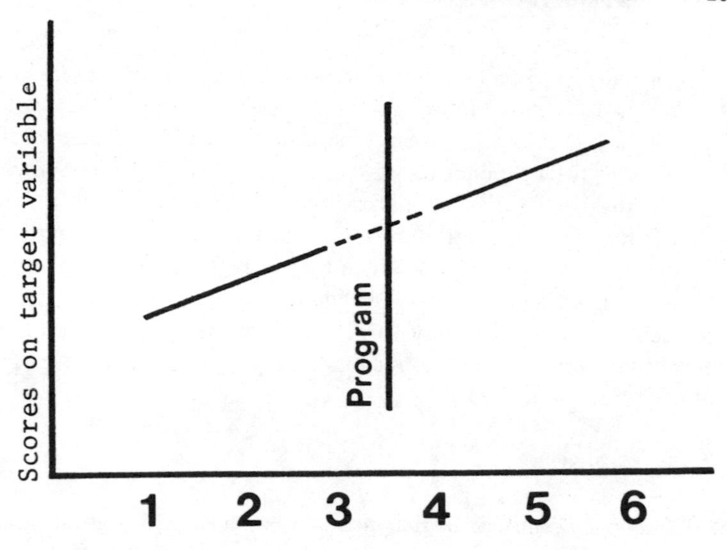

FIGURE 3
A TIME-SERIES INDICATING NO EFFECT ON A PROGRAM

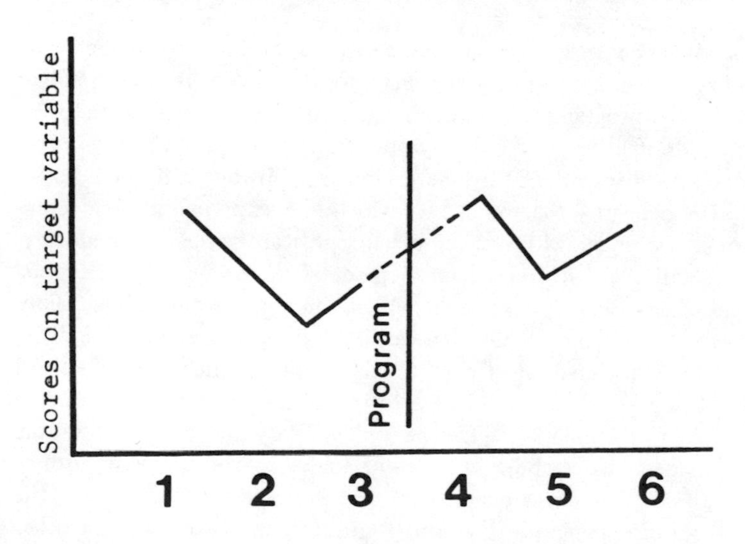

FIGURE 4
A TIME-SERIES INDICATING AN ILLUSORY EFFECT OF A PROGRAM

was going up at the same rate before the introduction and continues up at the same rate after the program's implementation.

The hypothetical data in Figure 4 are more problematic to interpret. The curve goes up from the introduction of the program to after its implementation. However, the wide variations before the introduction as well as those observed after the implementation provide no confidence concerning the effectiveness of the program. These are but two different types of findings that could be obtained from a Time-Series design. They do, however, demonstrate that such designs without any form of comparison group provide only rough evidence concerning the effectiveness of a program, irrespective of the number of measures employed.

CONTROL-SERIES DESIGNS

It has already been pointed out that procedures for matching groups might be vulnerable when data concerning significant confounding factors are unavailable. However, when data are available, nonequivalent comparison groups when used in Time-Series designs provide more reliable evidence on program effectiveness. Such designs are referred to as Control-Series designs. These designs attempt to control over those aspects of history, maturation, and measurement-remeasurement effects shared by the experimental as well as the comparison group.

The Control-Series design is especially suited for evaluating programs that are not being implemented simultaneously to all of the units in a target population. Many programs are matters of state and/or local jurisdiction rather than national. In such cases opportunities are available to assess program accountability with Control-Series designs. Obviously, in such cases a major problem concerns the extent to which the experimental and comparison groups are equal with respect to all significant variables. Obviously, units such as cities are not equal in every detail. Consequently, the issue becomes one of assessing the extent to which "an imperfect match plausibly explains away the outcome without necessitating the assumption that [the program] under study had an effect" (Riecken and Boruch, 1974:99). All other things being equal, the validity of a causal inference will increase the less likelihood for dissimilarities. Thus, supplementary and/or circumstantial evidence are needed to establish the extent to which the compared groups were similar. Factors such as changes in data collection procedures in the experimental group, changes in the construction of the measures, and other changes unique to the experimental group might lead to incorrect inferences concerning the accountability of a program.

REGRESSION-DISCONTINUITY DESIGNS

Given the reality of scarce resources, some programs are not being applied to all the units of a target population but rather to a select few. This creates a situation where there are more experimental units available for a program than the number of those who are actually going to be inducted into a program. A similar condition occurs for pilot programs where a new program is tried out on a limited number of cases.

The underlying idea of the Regression-Discontinuity design is that if a condition emerges where there are more target units than program space, some of the eligible units can be used as a comparison group. Under such circumstances, the most effective strategy would be the random assignment of target units occurred across the full range of eligibility. By keeping records on the randomly selected comparison group, one could, at later points of time, measure the target variable(s) in both the experimental and comparison group(s). In practice, however, randomization procedures are rarely employed. They are substituted for political reasons by criteria such as "the most needy," "the most deserving," or "first come." In such cases tie-breaking randomization procedures can be used to control for plausible rival hypotheses. The idea of tie-breaking experimentation evolved from the Thistlethwaite and Campbell (1960) experiments on the effects of receiving fellowships; that is, do academic awards exert an effect on later success in life? The postulated relationship is illustrated in Figure 5 where preaward ability and merit is measured on the horizontal axis and later achievement on the vertical. Students obtaining higher scores on the pre-test are most

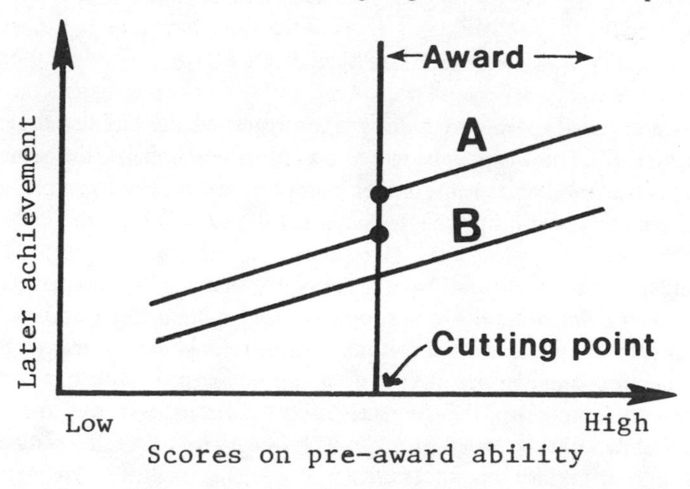

FIGURE 5
TIE-BREAKING EXPERIMENTATION AND REGRESSION DISCONTINUITY

deserving and in fact receive the award; they also do better in later achievement, but does the award exert a causal effect, given that the higher-ability students would generally have done better in later achievement anyway?

In this case randomization of the award is impossible in light of the policy to reward ability and merit. Nevertheless, Campbell (1969:420) suggests that "It might be possible to take a narrow band of ability at the cutting point, to regard all these persons as tied, and to assign half of them to awards, half to no awards, by means of tie-breaking randomization." If the tie-breakers would show an effect, there should be an abrupt discontinuity in the regression line. The discontinued regression line A in Figure 5 demonstrates a case in which higher pre-program scores would have led to higher post-program scores even without the application of the program, and in which there is in addition a substantial program effect. The regression line B shows a case where a program has no effect.

The essential requirement for this design is a sharp cutoff point on the eligibility criterion. The other requirement is that there be sufficient number of units. One way to maximize the number of units who were tied at the cut-off score is to use relatively large class intervals within which scores would be regarded as equal. The third requirement for the application of the Regression-Discontinuity design is that the eligibility criterion be quantified. For some public programs a quantitative criterion is already included among several eligibility criteria. Other programs, however, lack any quantitative criterion for determining eligibility. In such cases, if the design is to be applied, ranking or rating procedures can be employed (Nachmias, 1978).

STRUCTURAL EQUATION MODELS

In basic research pre-experimental designs are regarded the least credible for inferring causality. Too many intrinsic and extrinsic factors are not being controlled for; hence valid causal inferences cannot be made. For example, a very wanting form of pre-experimental design is as follows:

$$X \qquad 0_i$$

In this design some type of program is implemented and subsequently measurements are carried out. Such a design is most vulnerable. There are numerous rival hypotheses that could explain differentials in the measurements. First, measurements taken at the observation period would have no meaningful basis for comparison. Furthermore, such a design fails to provide any evidence as to whether a program had any effect on its target. To obtain such evidence it is necessary to add at least pre-program measures. History, maturation, and regression artifacts are also not being controlled for.

To compensate for the inherent weakness of pre-experimental designs, specialized data analysis techniques could be employed. Traditionally, such techniques have taken the form of multivariate data analysis where statistical controls (e.g., partial coefficients) are activated when physical control is either impossible or impractical. This approach is useful for eliminating certain rival hypotheses; statistical controls are preferable to no controls at all. Yet, for program accountability research, a more inclusive approach is needed to infer causal relations from pre-experimental designs. Structural equation models are aimed to serve this purpose.

An influential contributor to structural equation models is Simon (1957) who suggested restricting the notion of causality to simplified models that are subject only to few limitations and that would apply to the real world. One starts with a finite number of specified variables and constructs an explicit model from which testable hypotheses can be deduced. If, after testing, the model proves inadequate, additional variables can be introduced or the model can be modified. Underlying this process must be the understanding that "there is nothing absolute about any particular model, nor is it true that if two models make use of different variables, either one or the other must in some sense be 'wrong' " (Blalock, 1964:15).

Having delineated the variables to be included in the impact model, one can write a separate equation for each variable as a possible target variable. Assuming that each of the variables in the model is postulated to be caused by all of the remaining variables, and denoting the variables as $X_1, X_2, \ldots X_k$, the following set of regression equations can be written:

$$X_1 = a_1 + b_{12}X_2 + b_{13}X_3 + \ldots + b_{1k}X_k + U_1$$

$$X_2 = a_2 + b_{21}X_1 + b_{23}X_3 + \ldots + b_{2k}X_k + U_2$$

$$\vdots$$

$$X_k + a_k + b_{k1}X_1 + b_{k2}X_2 + \ldots + b_{k,k-1}X_{k-1} + U_k$$

The U_i's are "disturbance terms." These refer to the effects of exogenous variables (variables not explicitly defined in the impact model) on the appropriate target variables. If the value of each of the program variables in the system of equations is determined independently of the k target variable, the set of equations can be estimated by calculating k single equation ordinary least-square regressions. However, if one or more of the k target variables are also program variables in one or more of the equations, the system of equations is termed a "simultaneous system." The parameters of structural equations in a simultaneous system of equations cannot be estimated by ordinary least-square regressions since if, say, X_1 is estimated by X_2, the regression coefficient of the ordinary least-square regression of X_2 on X_1 is biased. One procedure for obtaining estimates of the parameters of structural

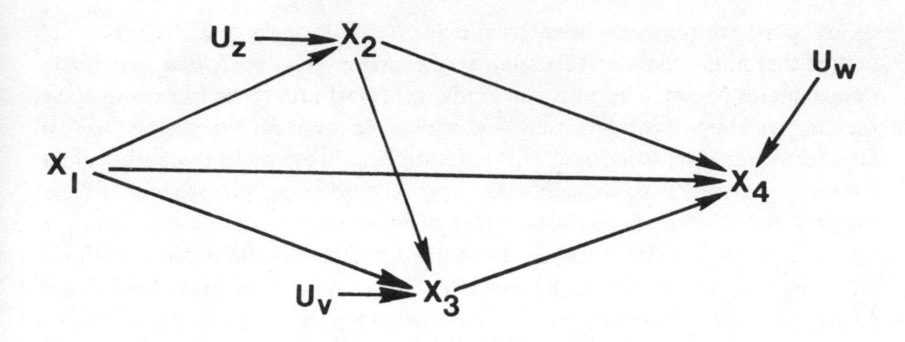

FIGURE 6
A FOUR-VARIABLE IMPACT MODEL

equations in a simultaneous system is the two-stage least-squares (Duncan, 1975:95-99).

In addition to an overall estimation of the model's adequacy, structural equations allow the assessment of the direct and indirect effects that one variable has upon another. For example, in Figure 6 it is observed that X_1 has a direct effect on X_4; X_1, however, exerts also indirect effects on X_4 via its effects on X_2 and X_3. Thus one can calculate the magnitude of the direct and indirect effects which helps in understanding the operative causal mechanisms of public programs. It is beyond the scope of the present paper to elaborate on the technical issues involved with structural equation models. Recent works (Duncan, 1975; Heise, 1975) in quantitative data analysis elaborate on these issues. Essentially, I attempted to convey the potential of structural equation models for program accountability research in pre-experimental settings.

A CONCLUDING REMARK

The broadening concept of public accountability to include concerns over program accountability has not resulted in more effective utilization of research information in the policy making process. A major reason for this failure is that program accountability research has not produced a body of systematic, general and accurate information. Given that at the heart of program accountability is the notion of causality, useful information for policy makers should demonstrate the extent to which public programs are causally linked to target variables. This inevitably leads to problems of research design.

My purpose here has been to provide a codification and discussion of factors that might threaten the validity of causal inferences. Taking the classic experimental design as the most credible model of proof, the paper surveyed the major design strategies that are applicable to program accountability research. Some would argue that experimental designs provide the most convincing evidence concerning the causal effects of public programs: "the capacity to produce a particular outcome deliberately in a randomly assigned group of persons is the surest testimony of the effectiveness of the program" (Riecken and Boruch, 1974:115). Indeed, when randomization is feasible, the optimal research design is the experimental one. However, social, political, and practical considerations may impede on the application of experimental designs. In such cases causal inferences are considerably more difficult to establish. Yet, quasi-experimental designs and structural equation models provide statistical approximations to experimental designs, and in this sense they are preferable to nonstructured research. Furthermore, the more systematic and rigorous the research design, the more valid and useful would be the information obtained for assessing the accountability of public programs.

REFERENCES

BERNSTEIN, I. (1976). Validity issues in evaluation research. Sage contemporary social science issues, 23. Beverly Hills, Ca.: Sage Publications.

BLALOCK, H. (1964). Causal inferences in nonexperimental research. Chapel Hill: University of North Carolina Press.

BOTEIM, B. (1965). "The Manhattan bail project: Its impact in criminology and the criminal law process." Texas Law Review, 43(February):319-331.

CAMPBELL, D. (1969). "Reforms as experiments." American Psychologist, 24(April): 409-429.

--- and STANLEY, J. (1966). Experimental and quasi-experimental designs for research. Chicago: Rand-McNally.

CHAPIN, F. S. (1940). "An experiment on the social effects of good housing." American Sociological Review, 5(December):868-879.

COCHRAN, W., and COX, G. (1957). Experimental designs (2nd ed.). New York: John Wiley.

DUNCAN, O. (1975). Introduction to structural equation models. New York: Academic Press.

FISHER, R. (1937). The design of experiments. London: Oliver & Boyd.

FRANCIS, G. (1961). The rhetoric of science. Minneapolis: University of Minnesota Press.

FREEMAN, H. (1977). "The present status of evaluation research." Pp. 22-39 in M. Guttentag (ed.), Evaluation studies: Review annual (vol. 2). Beverly Hills, Ca.: Sage Publications.

HEISE, D. (1975). Causal analysis. New York: Wiley-Interscience.

HORST, P., et al. (1976). "Program management and the federal evaluator." Public Administration Review, 36(July/August):220-235.

HOUSTON, T. (1972). "The behavioral sciences impact-effectiveness model." Pp. 51-65 in P. H. Rossi and W. Williams (eds.), Evaluating social programs. New York: Seminar Press.

LANA, R. (1969). "Pretest sensitization." Pp. 119-141 in R. Rosenthal and R. Rosnow (eds.), Artifact in behavioral research. New York: Academic Press.

NACHMIAS, D. (1978). Public policy evaluation. New York: St. Martin's Press.

NUNNALLY, J. (1975). The study of change in evaluation research: Principles concerning measurement experimental design, and analysis." Pp. 101-137 in E. Struening and M. Guttentag (eds.), Handbook of evaluation research (vol. 1). Beverly Hills, Ca.: Sage Publications.

PHILIPS, D. (1971). Knowledge from what. Chicago: Rand-McNally.

RIECKEN, H., and BORUCH, R. (eds.) (1974). Social experimentation: A method for planning and evaluating social intervention. New York: Academic Press.

SIMON, H. (1957). Models of Man. New York: John Wiley.

SOLOMON, R. (1949). "An extension of control group design." Psychological Bulletin, 46(January):137-150.

THISTLETHWAITE, D., and CAMPBELL, D. (1960). "Regression-discontinuity analysis: An alternative to the ex post facto experiment." Journal of Educational Psychology, 51(December):309-317.

13

Evaluation and Accountability

THOMAS E. JAMES, JR.
RONALD D. HEDLUND

□ IN RECENT YEARS, increasing demands have been made for government-sponsored services and programs designed to alleviate the problems facing our society. At the same time, resources have become scarcer and an increasingly aware public has demanded more governmental accountability in the use of those resources. Massive programs have been designed and resources allocated with little, if any, evidence available to indicate that any positive change is taking place because of these programs. Are these programs working or should government get out of the social service business? Should these resources (tax dollars) be spent on alternative programs which are more likely to be successful in addressing society's problems? What are the effects of government programs–negative and positive, intended and unintended? What are the long-term consequences, as well as the short-term effects, on those people touched by program activities? Do poorly designed and/or administered programs exist with potentially damaging outcomes for the very people they are intended to serve? The fundamental basis of our democratic society requires that those entrusted with public resources and responsible for applying them answer these questions and render a full accounting of their activities.

Public accountability demands that social programs be evaluated in order to find out what works and what does not work, and to find the best match between available resources and the programs most likely to have the greatest benefits. Public officials and program administrators are being asked to justify the need for programs and services, as well as the success these programs have in meeting needs. Further, they are asked to seek better ways of doing things while, at the same time, providing for the efficient management of programs.

In order to do this, political leaders, public administrators, and citizens need as much information as possible about the choices available and the consequences of each choice. This need stimulated the development and use of policy, program, and agency evaluations as a means for providing more complete and objective information on the costs, products, and effects of social programs. Evaluations can be used as a management tool to enhance effective management and the allocation of limited resources, as well as a means for promoting accountability.

The federal government has shown a heightened interest in the evaluation of its programs and evaluations have increased in both the executive and legislative branches. On the executive side, the Evaluation and Program Implementation Division of the Office of Management and Budget (OMB) was formed in 1974 as a focal point for activities to improve evaluation as a management tool. On the legislative side, the General Accounting Office (GAO) is engaging in its own audit and evaluation of executive agency programs. The GAO has had broad responsibilities for the review, evaluation, analysis, and audit function for several years. The Congressional Budget and Impoundment Control Act of 1974 strengthened congressional emphasis on evaluation and called for improved methods of evaluation. The act requires, among other things, that "the Comptroller General shall develop and recommend to the Congress methods for review and evaluation of government programs carried on under existing law" (U.S. General Accounting Office, 1975:i).

Legislation often includes provisions for evaluation, including accountability to federal level authorities as well as to local consumers. For example, in 1975 Congress passed the Community Mental Health Centers (CMHC) Amendments that tied continuation of the program to more tightly reined program administration and program evaluation at both federal and center levels. The Amendments call for

> an effective procedure for developing, compiling, evaluating, and reporting to the Secretary statistics, and other information relating to (1) the cost of the center's operation, (2) the patterns of utilization of its services, (3) the availability, accessibility, and acceptability of its services, (4) the impact of its services upon the mental health of the residents of its catchment area. . . . [further] such community mental health center will, in consultation with the residents of its catchment area, review its program of services and the statistics and other information referred to [in the preceding statement] to assure that its services are responsive to the needs of the residents of the catchment area. [Windle and Ochberg, 1975: 32]

Provisions of this nature in federal, state, and local legislation are likely to increase. This does not mean that it is necessary for program and project directors to be experts in evaluation; but it does mean that they must be aware of requirements for evaluation, and they must be ready to implement effective evaluations to help assure accountability. In order to accomplish this, public officials, program directors, and citizens, must be educated so that they are aware of the factors to be considered in an evaluation and questions to be answered when becoming involved in an evaluation effort. The following pages will discuss some of these questions in an attempt to focus attention on the issues surrounding the design of an appropriate evaluation methodology.

WHO IS THE AUDIENCE FOR WHICH THE EVALUATION IS BEING CONDUCTED?

It is important to recognize that varying viewpoints and priorities can exist among those interested in the evaluation. The official sponsor may be a program advisory committee, but the actual users may be the committee staff, the program director, or even directors of related programs. The purpose(s) or function(s) to be served by the evaluation will be dependent upon the *needs* of the *users* of the information produced. Different people may have different needs, depending on their program role. Those who funded the program may be interested in financial management and overall efficiency, while the program users or clients may focus on the adequacy of the services provided to meet their needs. Those charged with administration and direction of a program will most likely be concerned with both aspects.

Evaluators may not know in advance who *all* the potential users will be, but they must know who will be the primary initial users in order to select an appropriate evaluational design. Even if all the users are identified, it may not be possible to design. Even if all the users are identified, it may not be possible to design an evaluation which will meet all the varied needs due to constraints like time and money. Those agencies, groups, or individuals to whom the program is most accountable, not only help determine the priorities and goals of the program, but also the purposes for which an evaluation is to be conducted.

In addition to differences in viewpoints among those who will use the evaluation information, differences in basic orientation can exist between the user and the evaluator, especially if the evaluator has an academic background. The user is concerned with the specific application of knowledge, stressing the solution of immediate problems. The evaluator might be more interested in the acquisition of knowledge for its own sake or more concerned with long-term problem-solving. The evaluator focuses on the desire for generalizations, while the users emphasize the uniqueness of their own particular programs. These differences are most likely to cause problems when the purposes of the evaluator and user are at odds. For example, in order to execute an experimental design, the evaluator may want a client to be assigned randomly to a control or treatment group. From the viewpoint of program personnel, it may be preferable that all clients have the opportunity to benefit from the program's services and, hence, they should all be in the treatment group. The evaluator may have to adjust his research design in order to have access to a real world laboratory in which to test his theories and the practitioner may have to modify his goals in order to have valid, reliable, and precise data for evaluation purposes.

WHAT IS THE PURPOSE OR FUNCTION SERVED BY THE EVALUATION?

Two prerequisites for obtaining useful results from an evaluation are clarity and specificity regarding the study's purpose. A number of program aspects can be examined, depending on the specific needs of the audience: product, process, impact, efficiency, and compliance.

Product. The purpose of a product evaluation is to review the quantity and type of products or services produced by a program. For example, how many clients were processed by a program? How many persons were placed? How many persons completed a remedial reading program? How many houses were rehabilitated? Assessment of a program's products takes place without regard to outcomes (effect or impact); however, in order for an evaluation to study a program's product, the goals of the program must be product or service-delivery oriented.

Process. A process evaluation seeks to answer the "how" questions related to the program; how was the program operated and how were the clients served? This serves the function of delineating how the program was organized and administered; it is concerned with such things as organization structure and competence, administrative style, internal communications networks, management information systems, and modes of intervention. Process evaluations are basically descriptive and diagnostic; they attempt to find out what ways of doing things work and what activities lead to effective and efficient programs. A process evaluation may be an end in itself or it may be a part of some more comprehensive evaluation.

Impact. Impact evaluations consider the effects of program activities on intended and unintended clients; have beneficial outcomes been achieved? The *results* of program activities are measured, as opposed to the activities themselves. Rather than just documenting the number of placements for a job referral service, an impact evaluation might go on to examine the types of jobs in which people were placed, retention rates, job satisfaction, and the effect of the job on economic and psychological well-being. An impact evaluation would focus on the degree to which children's reading and comprehension scores increased and what happened as a result of this, and not just on how many students completed the reading program. In terms of the housing example mentioned earlier, a number of questions would be explored: What was the quality of the work done during the rehabilitations? How much and what kind of inconvenience did the homeowners have to endure? Were the homeowners satisfied with the work? Were the necessary repairs made? Did the repairs improve the living environment? Is the rehabilitated house worth more, or is it easier to sell?

Just as a specific statement of purpose is needed in order to design an evaluation that will produce useful information, the evaluator needs a clear, definitive statement of intended program results so that appropriate evidence about impact can be gathered. When program intent is vague or ambiguous, assessing impact becomes more difficult because the evaluator may be unable to identify where to look for a program's effects. Further, it becomes more difficult to identify unanticipated and/or negative consequences of a program as well.

Efficiency. A common concern of those funding social programs is with a program's efficiency–for example, its cost-benefit ratio. Could a comparable level of service or benefits be achieved at lower costs? For the same costs, could greater services be realized? A number of cost-analytic designs are available for assessing program efficiency: cost accounting, cost-benefit analysis, and cost-effectiveness analysis. All of these emphasize a relationship between the cost of operating the program and accomplishing program objectives and providing services, outputs, benefits, and effects. Data gathered to document efficiency provide decision makers with information for choosing among alternative programs based upon their relative costs.

Compliance. The purpose of compliance, control, or monitoring is to ascertain the degree to which the program is administered in accordance with the rules, regulations, and other guidelines governing such a program. Both administrative and financial compliance might be examined. Questions might include: What are the program's policies, are they appropriate, how is adherence to policy promoted? Is the program administered in accordance with relevant rules and regulations external to the program? Are the staff/client and professional/supporting staff ratios consistent with relevant guidelines? What are the procedures for verifying client eligibility? How have program resources been allocated and expended? What are the program's financial accounting procedures? Are proper reporting procedures being followed? This is by far the most common form of evaluation.

In many instances, decision makers have a need for more than one type of information. Product, process, impact, efficiency, and compliance evaluations are all interrelated and consideration of one aspect often leads naturally into another area. Nevertheless, the specific evaluational design methodology used is partially dependent on the function(s) to be served. Evaluation sponsors should be specific about the reason for conducting an evaluation in order to facilitate the design of an appropriate evaluation methodology.

WHEN SHOULD THE EVALUATION BE CONDUCTED?

An important consideration relates to when the evaluation should be initiated. Some have argued for beginning an evaluation after a program is implemented and has had an opportunity to operate for a period of time. Others argue that an evaluation must go hand-in-hand with program inception and planning. To some extent, the timing of an evaluation is directly related to the purpose for which it is being conducted, but an evaluation almost always meets its goals more effectively if consideration is given to it when a program is being planned and implemented.

Related to this is a concern regarding when evaluational reports should be provided—at points during the program's operation; or at the end, after a program is completed. Many decision makers prefer periodic reports during a program's existence so that adjustments can be made and ineffective programs terminated. On the other hand, a premature evaluation may kill a program before it has had a chance. Further, periodic evaluations make the evaluation itself a component of the program, thus confounding program effects. For example, programs that are intended to be highly innovative are likely to have a large number of unanticipated problems requiring modification or refinement in the program itself. Formative evaluations—those designed to provide ongoing feedback for program redesign and refinement—will detect these problems and provide information for corrective action (Scriven, 1967). Such evaluations increase the likelihood of developing a successful program. However, if the evaluation employs an experimental design, an interim report might prompt a program change by administrators that would prohibit the complete execution of the design. A summative evaluation, on the other hand, focuses on a finished program of activity with the aid of providing information to answer these questions: was it worth it, and should it be continued? This type of evaluation can examine the program as a whole, or look at separate components of the program. The focus is on the degree to which a program has been successful in accomplishing its goals and objectives (Scriven, 1967). With a highly innovative program that is undergoing a lot of change, a summative evaluation would not be appropriate until the program has reached a relatively stable state; otherwise, a negative evaluation might result at a time when the positive benefits have not had an opportunity to develop.

Too often, little attention is given to the long-range effects of program activities. Some effects (both positive and negative) may not become manifest until months or years after the initial evaluation period; other benefits may be short-lived and the evaluation might give a positively biased picture of the situation. Consequently, consideration should be given to follow-up measurements of program effects.

WHO SHOULD CONDUCT THE EVALUATION?

Those requesting an evaluation will need to consider the advantages and disadvantages of conducting an "inside" versus an "outside" evaluation. An inside evaluation is one in which the evaluators are staff members of the program which is being evaluated. An outside evaluation is conducted by people who are not associated with the program. They may be private consultants, staff of a governmental evaluation unit, or members of the academic community.

Perhaps the biggest advantage of an outside evaluation is likelihood of more objectivity. Since those conducting the evaluation have no vested interest in the program, it is easier for them to maintain their neutrality. This does not mean that outside evaluators have no personal interest in a program's value or utility, but it does mean that such interests are much less likely to enter into the evaluation. Further, an outside evaluator is more likely to be accepted as an unbiased expert by those being evaluated and by various individuals and groups who will use the information collected. As a result, it is less likely that the findings will be disputed and that recommendations will be challenged.

Outside evaluators are not without their liabilities, however. In order for members of any program staff to talk freely about program activities, it will be necessary for the evaluator to gain their trust and confidence. Since the evaluator is an unknown intruder into the program, anxiety may increase, especially during the initial stages of the evaluation. Informational briefings explaining the evaluation and encouragement from program administrators can help reduce the time necessary to build trust and confidence, but this will remain a problem. Time is also necessary in order to become familiar with goals, policies, and operations of a program. Some of this information can be gleaned from program documents, but a certain amount of interaction with program staff will always be necessary. The more time spent by program staff dispensing information through conversations, meetings, and briefings, the less time they have to spend on the performance of their duties. For evaluations of small or relatively simple programs, the time requirements for access to staff will probably not be very great and the disruption of operations should be minimal.

Since an inside evaluator is already a member of the program staff, the problems of entry and the accumulation of background information are much less severe. Even if the evaluator does not know each staff member personally, the assumption is that access and trust building will be greatly facilitated since he is already an "insider." Being an insider, the staff evaluator has a vested interest in the success of the program, as well as the image it portrays. Consequently, the findings of the evaluation (especially if they are positive) are very likely to be open to questions of bias and the evaluator's ability to be objective.

There are times when an internal evaluation would be appropriate. For example, a formative evaluation which interacts with program design would be more amenable to the heavy involvement of an inside evaluator. This would pertain particularly when program processes are the main concern. A summative evaluation, involving an assessment of the overall worth of the program, would be more appropriately conducted by an external evaluator in order to maintain a maximum objectivity.

Cost considerations are sometimes a factor when deciding whether the evaluation should be conducted by an outside group or internal staff. On the surface, it appears that an inside evaluation would cost less. In terms of primary or direct costs, this might be the case. The staff is already available; and the facilities, materials and supplies, and support staff of the program are there to be used. Secondary or indirect costs often operate against the selection of an inside evaluator. If the evaluation is to be conducted properly to produce useful information, sufficient time must be devoted to its design, execution, and analysis. This can detract from the time the staff evaluator(s) has to devote to day-to-day activities and responsibilities–resulting in a possible loss of service to program clients. On the other hand, lack of sufficient time devoted to the evaluation effort will prolong the whole process (which can have a detrimental effect if the results are to be used as feedback for redesign and refinement) and possibly prevent its completion.

In addition to the time required away from regular responsibilities and the possible difficulty in maintaining objectivity, internal staff may not have the appropriate expertise to conduct a professional evaluation. Even if the expertise does exist within the program staff, other factors may influence the decision toward selecting an outside evaluator–e.g., conditions in the political environment may warrant an outside evaluator in order to prevent any charges of a biased evaluation.

The final selection should be based on the purpose of the evaluation as reflected by the needs of the program and the evaluation sponsors, and by the evaluation methodology to be employed. Whenever specialized knowledge of design, data collection, or analysis is needed, an outside professional evaluator should be used. If factors such as budget constraints prohibit hiring an outsider, an evaluation consultant should be engaged to guide the efforts of the internal staff evaluator.

WHAT ARE SOME OF THE SITUATIONAL FACTORS THAT WILL IMPACT ON AN EVALUATION?

Political Context. One of the major constraints in conducting high quality and useful evaluations is the political context within which they are carried out. When an evaluation is viewed as an important part of the policy-making process and key officials support the evaluation, such support will enhance the evaluator's ability to conduct a rigorous and useful evaluation; however, when an evaluation is perceived to be a "necessary evil" in order to satisfy the requirements of rules and regulations, or is an afterthought in the policy-making process, the conditions may prohibit using the most appropriate methodology. The result may be the execution of an evaluation design that will produce information of marginal utility to the decision makers.

The evaluator who is not sensitive to the relationship of evaluation to the political environment is destined to be frustrated and disillusioned. Carol Weiss (1973) notes three ways in which political considerations intrude on the evaluation. First, the programs to be evaluated are the result of political decisions. Programs are part of the political process, from initial proposal to implementation. They emerge from the arena of political support, opposition, and bargaining and are not neutral and antiseptic entities. Consequently, the program administrator is likely to have a different view of the system and its important components than is the evaluator. At times, the political sensitivities of program administrators may inhibit their willingness to undertake an evaluation; and if an evaluation is conducted, their cooperation may be limited as is their attention to the findings and recommendations.

Second, evaluation is one phase of policy analysis. The intention is to generate information for the policy- and decision-making process. Thus, whether or not the evaluator intends it, evaluation reports enter the political arena. This does not mean that policy makers rely only on data from evaluations for making decisions about social programs, but it does mean that evaluations become a part of the decision-making process. Once in the arena, evaluation results must compete for attention with all the other sources that are viewed as important in the decision-making process. Usually, the results of a program evaluation are less important than the feelings of program supporters and detractors (especially if they carry political clout), the shape of the federal budget, the program's congruence with prevailing values, the views of voters, or the program use as payment of a political debt. One of the reasons that evaluation results are often ignored is that the evaluation is addressed to the official goals of the program, rather than to the political goals of important decision makers. Those who are concerned with official program goals are more likely to pay attention to and use evaluation results. While evaluations are not the only source of information in the democratic policy process, they can be a source of unbiased and useful information.

And third, evaluation itself has a political orientation. Inherently any evaluation produces information about program operations, goals, and outcomes. For example, an evaluation which seeks to determine how successful a program has been in achieving its goals, assumes implicity: (1) the desirability of achieving those goals; (2) that the program design is a reasonable way to deal with the problem; and (3) that the program has a realistic chance of reaching its goals. For some programs, social science knowledge and theory indicates that the program goals and problem diagnosis were not well reasoned, that the selected intervention was inappropriate, or that the outlook is not good for achieving the program's goals. Yet evaluations usually proceed without considering the consequences of these assumptions. Another political statement is made when some programs are selected for evaluation and others are not. It is generally the new and/or innovative program that is subjected to detailed and rigorous examination. The older, more established programs continue, unanalyzed and untouchable, even if they are not achieving their goals.

The structure of the evaluation also has political overtones. Evaluations are usually sponsored by a "higher order" agency responsible for a program and not by the program's clients. Thus, the needs and concerns of clients may not surface during the design or evaluation stages. In addition, the evaluation reports go to the program administrators and sponsors initially and not to program clients. If the results are negative, the report may not be released at all, or released with an explanation that the evaluation did not measure the program's true impact or did not reflect the difficulties under which the program operates. Only by being constantly aware of the political context and learning how to operate within it, can evaluators hope to play a useful and constructive role.

Cost. Another factor that is important when considering an evaluation is the cost. As mentioned earlier, both primary and secondary costs are involved. Primary costs are those directly associated with the execution of the evaluation—manpower, physical resources, facilities, and so on. Secondary costs reflect the effects of the evaluation on program operations in terms of time commitments and other resources required to facilitate the evaluation—the more secondary resources used, the more the evaluation may detract from an agency's ability to provide other services and meet its goals. Long-term payoffs in terms of a more efficient and effective program may outweigh the short-term losses and inconveniences.

All programs do not require the same level of funding for evaluation. A number of factors can enter into the cost or conducting an evaluation. These are examples:

Type and stage of program development: A new or innovative program using an experimental design will require more money than an established program because the latter evaluation may be very limited in scope and require little in the way of funding.

Size of program: In terms of a percentage of a program's budget, small programs tend to cost more to evaluate than larger programs.

Types of decisions required: Formative evaluations, designed to provide ongoing information for continual refinement, usually cost more than summative evaluations aimed at assessing program impact.

Probability that findings will be used: The greater the likelihood that program decisions will be influenced by the evaluation, the greater the justification for raising the spending level to cover the best design possible. [Wholey et al., 1970]

Since many of the costs associated with the execution of an evaluation are design dependent, estimation of the required funding level must be made while determining the research design to be used.

Timeliness. A third situational factor is the timeliness expected in the information provided by the evaluation. Those who will use the information may want it as input for the decision-making process. These decisions might involve mid-course correction of program activities or an assessment of whether or not a program should continue. However, many study methodologies may require more time than the user can afford. Consequently, timeliness of reporting requirements must be taken into consideration when designing the overall evaluation and when reaching agreement on what the study can be expected to produce within given time frames. It is important that the evaluation effort start in conjunction with the program planning. An evaluator's involvement from a program's outset will facilitate his/her understanding the intent and purpose of the program and developing an appropriate design. This also permits use of data collection routines which can begin at appropriate times, thus ensuring the availability of evidence needed to assess program effectiveness. In some cases, it may be necessary to sacrifice the desire for a comprehensive evaluation in the interest of timeliness in reporting.

WHAT ALTERNATIVE DESIGNS ARE AVAILABLE TO GUIDE THE EVALUATION?

The selection of an overall design to guide the execution of an evaluation cannot be made in isolation. No one design or technique is *the* best way to conduct an evaluation. Serious consideration should be given to the problem to be examined and the conditions surrounding it–the constraints of time, money, data requirements, political feasibility, and so on. All too often attempts are made to force what is being evaluated to fit a particular design.

Rather, evaluators should select the most appropriate design(s), given the purpose of the evaluation and the conditions under which it must be conducted.

A number of design options are available for use by an evaluator. These alternatives generally are grouped into three categories–experimental, quasi-experimental, or nonexperimental (pre-experimental)–according to the degree of control that can be exercised over the key elements of the evaluation.

Experimental Design. A true experimental design allows the evaluator to exercise a great deal of control over all aspects of program intervention–when it will take place, who will participate, and what conditions will prevail. Such control allows evidence to be gathered that will assist in assessing whether or not the activities of the program are causally related to whatever effects are detected.

When employing an experimental design, members of a target population are assigned randomly to treatment and control groups. The members of these two types of groups are expected to be similar in relevant characteristics. Criterion variables are defined that are indicative of program effects, and measurements are taken on these variables before and after program intervention. From the data collected through these measurements, estimates are made of the changes and effects that can be attributed to contact with the program.

The analytical rigor of experimental designs makes them a very useful and powerful tool for assessing a program's impact. However, this must be balanced against other factors which act as constraints to their application. One of the major obstacles in conducting a true experiment is the requirement that treatment and control groups exhibit similar characteristics. In its pure form this means randomly assigning clients to a treatment *or* a control group before the program starts. For ethical and practical reasons, program administrators are rarely willing to authorize randomization as a means for selecting those individuals who will receive program services. Another constraint on the use of experimental designs is the requirement that treatment and control conditions be held constant during the period of program intervention. Altering the program once it is in process detracts from the evaluator's ability to link impact and effect to program activities. Hence, experimental designs tend to inhibit modifications and refinements during the course of the program and would not be appropriate in those instances where evaluation is viewed as a way to facilitate the continual improvement of the program.

Quasi-experimental Designs. Quasi-experimental designs are employed when it is not possible to control all the relevant conditions (e.g., assignment of individuals to treatment or control groups) necessary in order to conduct a true experiment. Nevertheless, evaluators using these designs do strive to extend the logicl of causal hypothesis testing into an environment where complete control is not possible. In many cases, quasi-experimental designs have the advantage of being more feasible to implement than the rigorous and sophisticated experimental design.

A number of quasi-experimental designs are commonly employed. A time series design involves a series of measurements at periodic intervals before the program was implemented and continuing after the program's services were provided. Changes in data trends are taken as evidence that the activities of the program are responsible for the measured changes. However, caution must be used when interpreting time series data in order to allow for an appropriate time lag between exposure to the program and manifestation of the impact of the services. In addition, care must be taken to separate out any cyclical phenomena which might influence trends in the data. In order to protect against the possibility that some other event, going on at the same time as the program (history) led to the changes, it is useful to have similar measurements from a comparable group that did not partake in the program.

When it is not possible to control the random assignment of individuals to control and treatment groups, a nonrandom *comparison group* is often used. The comparison group is constructed by selecting people from among those not receiving program services, but who exhibit characteristics as similar as possible to those of the treatment group. Before and after measurements are taken for each group and the differences in the results between the two groups assessed in an attempt to discern (as in an experimental design) which differences in outcomes might be attributed to the intervention of the program. Even with success in matching characteristics, however, there remains a greater danger that, without random assignment, the observed results are attributable to nonprogram influences. Further, it is sometimes difficult to define the precise set of variables on which the groups should be matched.

Quasi-experimental designs in natural settings also run the risks of possible bias resulting from self-selection by participants. Comparisons between the treatment and comparison groups can be confounded by the possibility that those who self-selected themselves into the program are likely to differ systematically from those who did not. These differences may be in terms of interest, initiative, value placed on the service, and so on. Thus, the use of comparison groups does not solve all of the problems derived from the lack of randomization, but controlling for some explanation of effect is better than no controls at all.

Another type of quasi-experimental design entails comparing similar programs or strategies in terms of their outcomes. This type of design has several advantages. It can provide information about the effectiveness of various alternatives and reduce the need to rely on control groups, as well as the requirements of an experimental design. If widely representative projects can be included, it will also promote the generalization of the results, increase the likelihood of identifying exceptional programs, and offer an opportunity to study what is operationally different about those programs. Although a number of advantages can accrue, comparisons comprehensive enough to be useful are expensive and difficult to manage.

Nonexperimental Designs. One of the purposes of conducting an evaluation is the desire to be able to say, definitively, that a particular program or strategy led to a particular outcome which would not have occurred without the program intervention. When contextual constraints do not permit the use of experimental or quasi-experimental designs, evaluators must rely on nonexperimental designs: for example, a post-test only or a one-group pre-test/post-test. The major drawback stems from the fact that these designs do not provide much basis for the evaluator to make causal inferences regarding program effects. Nevertheless, if the data are collected systematically, they can provide more information than would have been available without the study.

The post-test only is the simplest nonexperimental design and also the one most lacking in control and value. With this design, measures are taken from a group after they have been exposed to a program and no explicit comparison is made with any other group. The difficulty arises in arguing that members of the group are somehow different after exposure than before. Some measures can be employed in an attempt to generate comparison data. For example, participants can be asked to recall their status before the program, as a way of getting some pre-exposure data; however, the problems associated with retrospective reporting require care in interpreting any change. If available, other official records can be used as a source of information about program participants before they received any services. Finally, participants can be asked about their experiences with the program in an effort to discern any effects those experiences had on the participants.

An improvement on the post-test only design is the one-group pre-test/post-test design. This type calls for measurements to be taken before and after a group is exposed to a program. If program administrators are interested in the process of implementation, measures also can be taken periodically throughout the group's participation in the program. The scores can be compared to see if any changes take place between the measurement points. Unfortunately, it is not possible to say which changes resulted from the efforts of the program and which were due to external influences. While case studies of this type may provide little knowledge that is generalizable to other programs, they can provide information to program administrators that will help them assess and refine the operations of their programs.

Economic Analysis. A special set of designs that, to varying degrees, applies some of the features of experimental, quasi-experimental and nonexperimental approaches includes various types of cost-analytic research: cost-accounting, cost-benefit analysis, and cost-effective analysis. Cost-analytic designs are concerned with the relationship of program costs to program outputs and outcomes. Most of these studies attempt to provide information pertaining to optimal relationships between costs and program activities, usually in terms of a desired level of output or benefit at minimum cost, or maximum level of output or benefit within fixed-cost limits. Cost is defined not only in terms of dollars but also in terms of such things as requirements for additional skilled manpower; time required for implementation; impact on present manpower requirements for additional training or restructured roles; capital expenditures for facilities, operations, maintenance, and the like.

Cost-accounting attempts to identify cost associated with program products or components and is frequently done in conjunction with some form of program monitoring. It is concerned with how much money is spent on what kinds of activities in relation to which program goals. Its logical extension leads to program budgeting techniques which assess alternative program activities in relation to cost and utility, as part of the planning process. Since cost-accounting is geared toward an assessment of program efficiency, it is difficult to implement if program objectives have not been specified or categories of unit cost are ambiguous.

Cost-benefit analysis is more concerned with the relative effectiveness of alternative programs and strategies in terms of their costs and benefits. Rather than just assessing the costs of program outputs, cost-benefit analysis determines the cost, in terms of all resources associated with the achievement of program objectives. In order to do this, the evaluator must be able to identify the total costs of resources used to achieve objectives, in terms of monetary units, and the total value of the objectives achieved, in terms of those same monetary units. Then an assessment can be made of a program's monetary resources expended in relation to the monetary units of benefits gained—the greater the benefit-to-cost ratio, the more effective the program. With this type of information, decision makers are able to compare alternative programs. When the information is used for planning how resources can be used to achieve goals and benefits, this approach is likely to advance the most cost-efficient program.

One of the major problems with cost-benefit analysis is the requirement for establishing monetary values to be assigned to program objectives. When the program objective involves a product whose market value can be established, few problems are encountered; however, this is rarely the case with governmental social programs. What is the total value to be placed on protecting children or feeding hungry people? Measuring psychic and social benefits is difficult. For example, while the economic benefits that go along with employment are one factor in determining the results from a governmental jobs program, cost-benefit analysis does not consider other benefits that are also important–e.g., the psychological well-being that accrues from being employed. Cost-benefit analysis can be of utility in the systematic examination of alternative courses of action and their implications, but only when social benefits are not deemed to be of more value than economic benefits.

Cost-effectiveness analysis is a modified version of cost-benefit analysis. The basic difference is with the measurement of benefits. Unlike cost-benefit analysis, this type of design does not translate the program outcomes (benefits) into economic or monetary units. Specific measurable objectives are stated (a 15% reduction in crime) and alternative program activities designed to accomplish the objective(s) are compared with respect to cost. Outcomes are analyzed in terms of the rates, direction, nature, and amount of change. The focus is on developing a program that will accomplish the objectives at a given cost. The elements of the program can vary, but the objectives to be accomplished are specific and fixed–including psychological and sociological change.

HOW IS AN EVALUATION DESIGN SELECTED?

The selection of the most appropriate design will be influenced by a number of factors–cost, time available, the political environment, the types of decisions to be made, the desired nature of the conclusions, and so on. One of the most important determinants to consider is the purpose for conducting the study and the use to which the data will be put. For example, experimental designs are very powerful for assessing the impact of program interventions and measuring the degree of goal achievement. They can also be used for process and product evaluations by examining alternative program strategies for accomplishing program goals relative to program output. The use of control groups allows the evaluator to express a great deal more confidence regarding the validity of the evaluation than is possible with other types of designs. However, experimental designs are not very useful when the purpose is to assess program compliance with internal and external rules, regulations, and guidelines.

Quasi-experimental designs are appropriate for the same purposes as those noted for experimental designs. While they do not have the same vigor or validity as experimental design, quasi-experimental designs have the advantage of being feasible in many situations when a true experimental design cannot be implemented. Nonexperimental designs are most appropriate when the purpose of the evaluation can be served by using a case study method. While each of the nonexperimental designs has some major drawbacks, information gathered with this type of design can help the administrator improve the operations of the program. Since it is very difficult to separate out influences from outside the program, nonexperimental designs are not very useful for assessing impact due to program activities. Cost-analytic designs are concerned with efficiency and should be used when evaluating the relationship of program costs to the accomplishment of program objectives, the provision of benefits, and the effect on program clients. Whichever design is selected, it is important for the choice to be based on the purpose for which the evaluation is being conducted. This will help ensure that the information produced is maximally useful.

WHAT DATA COLLECTION TECHNIQUES ARE AVAILABLE?

The evaluator must design and implement data collection mechanisms that will generate the evidence required in order to evaluate the various aspects of a program. A wide variety of data sources exist, and standard research techniques can be used to collect the data. The most widely used data-collection techniques involve observing behavior and asking questions. Observations can be conducted using an unstructured or a structured format. Unstructured observation tends to be less obtrusive and is particularly useful during exploratory phases of an evaluation; however, comparability across observations is likely to be less reliable. Structured observations use pre-coded categories to record information and tend to have a higher degree of precision and reliability. But observers using a structured approach are usually very visible to program staff and participants and may produce changes in normal behavior and interaction patterns.

A somewhat more obtrusive means for collecting information involves talking to people and asking questions. Interviews can be conducted in person or by telephone and can be structured or unstructured. Using an interviewing strategy will provide an opportunity for the evaluator to interact with program participants--clients and staff alike. Questionnaires are instruments that are completed by the respondent and can be administered directly or sent through the mail. As with interviewing, large numbers of respondents can be reached, but somewhat more economically. Further, when program goals are related to learning situations, tests are often administered as a way of asking questions to generate information about the degree to which the goals are being met. Some types of data are not amenable to these techniques and the evaluator most exercise caution not to apply these, or any other techniques, indiscriminately.

Since both the observational and questioning methods tend to make participants more aware of their role as "subjects," either technique could alter behavior and confound the evaluation effort. An alternative strategy is the use of already existing data as an unobtrusive means for collecting information about the characteristics and behavior of individuals, without heightening their awareness that measurements are being taken. Government statistics (e.g., Bureau of the Census, Department of Labor, Social Security Administration, local school districts) provide a wide variety of data which describe the conditions within a given program target area and the characteristics of potential clients within that area. Various program records–financial, minutes, client files, progress reports–offer another source of useful data. Agency data have the advantage of saving time and money since data specifically pertaining to the program do not have to be generated from "scratch." On the other hand, evaluators must be sensitive to the existence of possible bias. The data may be reflecting the organizational, professional, and individual interests of those who designed and implemented the record-keeping system. In addition, agency records are frequently inaccurate and incomplete. Even if care was taken when recording the data, it may not be appropriate for the needs of the evaluation.

The use of multiple data sources and data collection techniques can help alleviate some of the problems encountered when an evaluator relies on any one source or technique. For example, multiple techniques can help control the effects of reactive research situations. The effects of being under observation are well documented, as is the tendency to give socially desirable answers in response to a questionnaire. At the same time, much of the data available from government sources may be out of date, incomplete, or not appropriate for the geographic units of analysis. Following through with the housing example mentioned earlier, interviews could be conducted with owners of rehabilitated houses to assess satisfaction with the rehabilitation efforts (outcome/effect). Observations could be made before and after the rehabilitation takes place in order to document (perhaps with pictures) the work that was performed. Program records could be used as a source of information concerning any manpower training that took place as part of program activities (process) and as a source of information concerning the number of houses rehabilitated (output). Thus, using a variety of sources and techniques may help provide a more comprehensive picture of the activity being evaluated.

Whatever data sources and collection techniques are used, data should not be generated just for the sake of gathering data. Careful planning should ensure that the time, expense, and energy devoted to the data collection activity produces more than a mountain of impressive, useless information. The evaluator must determine what data are needed and then select the most appropriate sources and techniques to collect the data. Like the choice of a research design, the data collection strategy must be related to the purpose of the evaluation, the situational conditions under which the study must be conducted and, in addition, to the design being employed.

Increasingly, public officials, both elected and appointed, are being held accountable for the actions they take which have an impact on the lives and social well-being of their constituents. The federal government, as well as local service consumers, is seeking justifications for the allocation of resources—how are dollars being spent and for what purpose? Are individual program participants and society as a whole benefiting from the programs and services being provided by government? Would our resources be better spent on alternative solutions to the problems facing our society? The need to answer these questions, the competition for scarce resources, and active public scrutiny has led to the increasing development and use of evaluational methodologies. In order for government to design effective programs for responding to identified needs, the capacity must be developed to assess the effectiveness of current and prior efforts. This chapter has discussed some of the key issues to be considered when building this capacity and undertaking an evaluation effort to advance governmental accountability.

REFERENCES

SCRIVEN, M. (1967). "The methodology of evaluation." Pp. 38-83 in American Educational Research Association monograph series on Curriculum evaluation, perspectives of curriculum evaluation. Chicago: Rand-McNally.

U.S. General Accounting Office (1975). Evaluation and analysis to support decision making (Exposure draft). Washington, D.C.: Author.

WEISS, C. (1973). "Where politics and evaluation research meet." Evaluation, 1(3):37-45.

WHOLEY, J. S., et al., (1970). Federal evaluation policy. Washington, D.C.: The Urban Institute.

WINDLE, C., and OCHBERG, F. M. (1975). "Enhancing program evaluation in the community mental health centers program." Evaluation, 2(2):31-36.

14

Are Traditional Research Designs Responsive?

ROBERT K. YIN

☐ THE PURPOSE OF THIS PAPER is to raise certain issues regarding the applicability of traditional research designs—i.e., those based on experimental psychology (Campbell and Stanley, 1966)—to the study of the accountability of public agencies. The accountability of such agencies, whether to legislative bodies, to other public organizations, or in particular to the clients who use the agencies' services, has become of increasing concern in American public policy. To a certain extent, these agencies must serve their own bureaucratic needs in order to survive (Downs, 1967), but it has also become evident that new techniques are needed to insure that these agencies continue to serve the public interest (Lipsky, 1977; and Greer, 1977). Such techniques might include:

> (1) The passage of "sunset" laws (deLeon, 1977).
> (2) The development of closer ties between professional groups and the clients they serve (including doctor-patient, police-citizen, and teacher-student relationships; see Yin and Yates, 1975).
> (3) The development of a whole host of citizen feedback mechanisms—including the use of vouchers, citizen surveys, or citizen-dominated governing boards—in order to guide agency policies. [Yin, 1975]

AUTHOR'S NOTE: *Discussant's remarks presented at the Conference on Public Agency Accountability in an Urban Society, Urban Research Center, University of Wisconsin—Milwaukee, April 3-5, 1977.*

This paper addresses the problem of designing empirical studies of accountability. The studies may focus, for instance, on developing new techniques or evaluating specific policy interventions. Typically, a research objective is to determine the extent to which a police or school department, a state highway agency, or a federal bureau is responsive to the needs it purports to serve. A basic review of the possible research designs, stemming mainly from the traditional literature on experimental and quasi-experimental designs, has already been given in a previous paper (Nachmias, 1977). The paper includes a thorough discussion of the threats to validity, the preferred designs, and the ways of dealing with less than ideal experimental conditions. There are four issues, however, that must be raised in the application of these research designs to the study of accountability. These issues are:

(1) The appropriate *unit of analysis.*

(2) The *interrelationships among alternative outcomes.*

(3) The *complexity of interventions.*

(4) The *assessment of voluntary efforts.*

In the aggregate, the problems posed by the four issues suggest that traditional research designs may not be the best approach for studying accountability, and that alternative approaches may be necessary. In this sense, traditional research designs may not be responsive to the needs for studying accountability.

FOUR DESIGN PROBLEMS

UNITS OF ANALYSIS

Many studies of accountability will focus on specific practitioners and their relationships to clients. Thus, a study of doctors and their patients could reveal the ways in which concerns for professional status may have begun to infringe on the ability of a doctor to provide adequate time and ministration to a patient (Greer, 1977). For such studies, the appropriate unit of analysis may be the individual practitioner, the individual client, or the interpersonal relationship between the two. These are units of study that are easily incorporated into the traditional research designs, and it is no surprise that the traditional designs are well-suited for studies of health care and of classroom behavior in education (e.g., educational psychology).

The most prominent set of emerging issues on accountability, however, goes beyond the concern for specific interpersonal relationships. Most often, it is an agency or organizational unit whose policies are of concern; similarly, it is the work of a citizen-dominated governing board that is a frequent subject of a citizen feedback study. For instance, one wants to know why some police departments appear to serve their citizenries well, whereas other departments appear to be less responsive. As another example, there is the problem of determining why some neighborhoods receive better services than others. In such instances, the appropriate unit of analysis is a formal group or organizational unit. In such cases, the traditional research designs are less suited than when the unit of analysis is an individual.

First, organizations are not easily defined. Should a study of a police department include the role of oversight agencies or personnel performing police functions but located in different organizations? Any administrator knows that agency jurisdictions differ, and that resources are commonly hidden in a manner that deviates from formal organizational charts, budgetary documents, or personnel job descriptions. Thus, the police department in a county-city government may be quite differently constituted from a police department in a city-dominated government. Second, organizational responses are not as easy to define as an individual's performance on a medical or educational test. One part of an organization may be responding differently from another part, and the unit of analysis is thus not unitary. These two problems alone create difficulties in applying any procedures such as random sampling or random assignment for any study of organizational behavior. One possibility is for a study to encompass the entire universe of possible organizations, as in Walker's study (1969) of innovations by state legislatures or Kaufman's study (1976) of organizational mortality. Yet, for many studies of organizational behavior, in-depth analysis means that only a small set of organizations can be examined, and the traditional research designs do not provide sufficient guidance for selecting appropriate samples under these conditions.

INTERRELATIONSHIPS AMONG ALTERNATIVE OUTCOMES

Most accountability studies need to focus on more than one outcome. The substantive issues usually require, for instance, that responsiveness, equity, and efficiency be at least three of the outcomes that are examined. Traditional research designs, however, are based on the researcher's ability to focus on a single outcome. Typically, the crucial experiment in psychology isolates on a single, critical dependent variable. The experimental design may allow for the fact that different measures will be used to assess this variable, but the notion is that the data will converge and that the essential hypothesis will be tested.

In contrast, accountability studies often begin with the acknowledgment that several of the relevant outcomes are partially exclusive of each other. Equity cannot be attained without some loss in efficiency or responsiveness; responsiveness cannot be attained without some loss in efficiency. The research study will of course examine the factors that lead to each of these three (or other) outcomes; but the traditional research design provides little guidance for weighing the exclusivity of the outcomes, which may be the most critical part of the policy problem. Accountability studies must therefore often incorporate an arbitrary set of weights imposed by the researcher or leave the final judgment to the policy maker. In either case, the rigor of applying traditional research designs cannot be captured, and the presumed benefits from such designs may be lost.

COMPLEXITY OF INTERVENTIONS

There is a wide variety of known policy interventions that can be initiated to improve agency accountability. For instance, school voucher schemes can be designed so that students can attend those schools that most appear to meet their instructional needs. The assumption is that, with the voucher as a feedback mechanism, schools will gradually adopt more responsive programs; viewed simplistically, a school (or school district) that does not adapt will not attract a sufficient amount of vouchers in order to continue its operations. However, such an intervention as a voucher scheme is a very complex affair, and research on program implementation is just beginning to acknowledge that similarly designed programs will vary substantially in their implementation from site to site or from time to time at the same site (Berman and McLaughlin, 1974; and Yin et al., 1977).

The traditional research designs attempt to treat this complexity by enumerating a long list of independent variables. Interventions, though, are themselves an organizational process, and change agents who have dealt with such interventions will easily recognize that an intervention is more than a collection of independent variables. Recent organizational research, focusing on issues related to accountability, have therefore found an increasing need to use case studies as a methodological approach for capturing the complexity of an intervention. Indeed, the most insightful studies of organizational innovation are often based on well-documented case studies (e.g., House, 1974; Gross et al., 1971). But the case study approach, no matter how appropriate, is not actually considered an experimental design. One-shot case studies are regarded as pre-experimental designs, and for good reason (Campbell and Stanley, 1966). From the point of view of traditional research design criteria, one-shot case studies pose insurmountable problems for establishing internal or external validity. At the same time, the traditional research designs do not offer an alternative other than using a design in which a complex intervention can only be treated as a collection of independent variables.

ASSESSMENT OF VOLUNTARY EFFORTS

Many strategies for improving accountability involve citizen or client efforts. Clients may form advisory or governing boards, they may be encouraged to make complaints, they may be asked to use vouchers, or they may participate in any number of ways in agency operations (Yin, 1975). In most of these activities, citizens will be participating as volunteers. By definition, voluntary activities are not susceptible to easy manipulation by a policy maker or an evaluation researcher. Once again, this means that traditional research designs offer only limited assistance.[1]

First, voluntary activities will emerge at a time and place that are not necessarily of a policy maker's (or researcher's) choosing. There may be insufficient time to collect baseline data, an inadequate basis for defining experimental and control groups, and no way of preventing some voluntary activity from later arising in a "control" site. Second, voluntary activities typically incur few costs, and the costs of conducting the research project can be substantially greater than the funds needed to support the voluntary activity. (One need only compare the per capita costs of a residential survey with the typical expenses for voluntary activities—e.g., gas for driving, refreshments for meetings, paper and supplies, postage, and so on.) Thus, the conduct of a traditional research project might generate public criticism for its relative costs.

To the extent that accountability activities involve citizen voluntarism, then, another peculiar set of obstacles is posed in attempting to use traditional research designs. These obstacles are, of course, over and above the problems raised by the previous three issues.

ALTERNATIVE APPROACHES

The purpose of a research design is to serve as the plan for bringing evidence to bear to establish an assertion; a design is a model or logic of proof (Nachmias, 1977). The main suggestion from the preceding four points is that traditional research designs—i.e., those stemming primarily from the field of experimental psychology—may create insurmountable problems in being applied to studies of accountability. One solution is to continue to make incremental modifications and compromises, so that evaluation studies, for instance, provide the best possible empirical information, even if somewhat flawed. This incremental approach appears to be the dominant contemporary theme.

Experimental research designs, however, are not the only logic of proof. Journalists, lawyers, historians, and indeed clinical psychologists all deal with empirical matters and do not use what we have called the traditional research designs. These other crafts follow other paradigms for relating evidence to assertions, paradigms that are no less legitimate than the experimental research designs. Thus, a second possible solution is to develop an entirely new logic for investigating the subject of accountability. The specific logic may not presently exist, but a concerted effort might be well worthwhile; in the long run, such an activity may have a greater payoff than the most exhaustive set of incremental adjustments to traditional research designs, because such designs may simply be inappropriate for sufficiently dealing with accountability.

Where could such an effort to develop a new logic begin? A starting point could be the codification of existing research designs in political science research, because accountability is ultimately a topic concerning the core of political science—the relationship between government and citizen. Note that a typical array of political science studies is not limited to experimental research designs. A different logic has been used to conduct studies of checks and balances among the three branches of government or, as another example, the making of foreign policy. (In fact, if the standard evaluation research design—based on experimental research—were applied to a study of checks and balances, the study would probably conclude that the checks and balances system is more costly and time-consuming and also arrives at poorer decisions because so many political compromises have to be made, and yet the study would have ignored what we value most about the checks and balances system.) The implicit designs are not easy to describe. Thus, a codification effort would help to identify the implicit designs and distinguish the more powerful designs from the weaker designs, just as Campbell and Stanley (1966) have done for experimental psychology. Political science, in short, needs its own Campbell and Stanley—or, to compare it to another field, its own historiography. This is indeed the real challenge for developing appropriate research designs to study the topic of accountability.

NOTE

1. A similar issue concerning the evaluation of voluntary crime prevention activities has been discussed in a separate paper (Yin, 1977).

REFERENCES

BERMAN, P., and McLAUGHLIN, M. (1974). Federal programs supporting educational change (Vol. 2), R-1859/2-HEW, September. Santa Monica, Cal.: The Rand Corporation.

CAMPBELL, D. T., and STANLEY, J. (1966). Experimental and quasi-experimental designs for research. Chicago, Ill.: Rand-McNally.

deLEON, P. (1977). The sun also sets: The evaluation of public policy, P-5826, March. Santa Monica, Cal.: The Rand Corporation.

DOWNS, A. (1967). Inside bureaucracy. Boston, Mass.: Little, Brown.

GREER, S. (1977). "On studying accountability of public agencies in an urban society." Background paper for the Conference on Public Agency Accountability in an Urban Society, Urban Research Center, University of Wisconsin–Milwaukee, April 3-5.

GROSS, N., GIACQUINTA, J. B., and VOGEL, M. E. (1971). Implementing organizational innovations: A sociological analysis of planned educational change. New York: Basic Books.

HOUSE, E. R. (1974). The politics of educational innovation. Berkeley, Cal.: McCutchan Publishing.

KAUFMAN, H. (1976). Are government organizations immortal? Washington, D.C.: The Brookings Institution.

LIPSKY, M. (1977). "The assault on human services: Street-level bureaucracy, accountability and the fiscal crisis." Paper presented at the Conference on Public Agency Accountability in an Urban Society, Urban Research Center, University of Wisconsin–Milwaukee, April 3-5.

NACHMIAS, D. (1977). "Design strategies for accountability research." Paper presented at the Conference on Public Accountability in an Urban Society, Urban Research Center, University of Wisconsin–Milwaukee, April 3-5.

WALKER, J. L. (1969). "The diffusion of innovations among the American states." American Political Science Review, 63(September):880-899.

YIN, R. K. (1977). "Evaluating citizen crime prevention programs." Paper presented at the National Conference on Criminal Justice Evaluation, Washington, D.C., February 22-24; also The Rand Corporation, P-5853.

--- (1975). "Citizen feedback mechanisms in municipal services." Paper presented at the Neighborhood Concepts of Local Government Seminar, Fort Wayne, Indiana, September; also The Rand Corporation, P-5553.

---, and YATES, D. (1975). Street-level governments: Assessing decentralization and urban services. Lexington, Mass.: Lexington Books.

YIN, R. K., HEALD, K. A., and VOGEL, M. E. (1977). Tinkering with the system: Technological innovations in state and local services. Lexington, Mass.: Lexington Books.

THE AUTHORS

MICHAEL A. BAER is Associate Professor and Chairman of the Department of Political Science at the University of Kentucky. He has written several works on interest groups including his coauthored book, *Lobbying: Interaction and Influence in American State Legislatures.* In addition he has published several articles on local methods. He has also produced several series of videotapes introducing social scientists to computer systems.

JAMES L. GIBSON is Assistant Professor of Political Science at the University of Wisconsin-Milwaukee. His primary research interest is in the politics of decision-making, and his published work has appeared in the *American Political Science Review,* the *Journal of Politics,* and *Law and Society Review.*

VIRGINIA GRAY is Associate Professor of Political Science at the University of Minnesota. She was coeditor of *Political Issues in U.S. Population Policy* and has published articles in *American Political Science Review, American Journal of Political Science, Polity, Policy Studies Journal* as well as in edited collections. Currently she is a Guest Scholar at the Brookings Institution where she is studying intergovernmental relations.

SCOTT GREER is a Professor of Sociology and Urban Affairs at the University of Wisconsin-Milwaukee. He has been Director of the Center for Metropolitan Studies at Northwestern University and the Doctoral Program in Urban Social Institutions of the University of Wisconsin-Milwaukee. His publications include *Social Organization, The Emerging City, Governing the Metropolis, The Logic of Social Inquiry,* and *Neighborhood and Ghetto* (with Ann Greer).

KEITH E. HAMM currently is an Assistant Professor in the Department of Political Science, Texas A & M University. In addition to his dissertation on "The Effects of Organized Demand Patterns on State Legislative Policy Making: The Case of Wisconsin," he has coauthored several articles and papers on measuring legislative performance and its effects on public policy-making. His current interests include research on environmental and energy policies.

RONALD D. HEDLUND is Professor of Political Science at the University of Wisconsin-Milwaukee and Center Scientist at the Urban Research Center. He is coauthor of *Representatives and Represented* and *The Job of the Wisconsin Legislator* and has contributed numerous articles to political science journals. In addition, he currently serves on the editorial boards of *American Politics Quarterly* and *Legislative Studies Quarterly.* Currently, his research is focusing on a cross-state study of legislative performance and policy-making.

THOMAS E. JAMES, Jr. is the Director of the WIN Evaluation Project and Associate Director of the CETA Review Project at the University of Wisconsin-Milwaukee. He received his Ph.D. in Political Science from the Ohio State University where he was Co-Director of Policy Action Clinic at the Mershon Center for Programs in Research and Education in Leadership and Public Policy. He is primarily interested in the promotion of knowledge utilization systems to facilitate the application of social science knowledge and methodologies to the formation, execution, and evaluation of public policies. In particular, his research interests focus on needs assessments and evaluation and he has conducted numerous studies in both areas.

JAMES H. KUKLINSKI completed his Ph.D. at the University of Iowa. He is currently Assistant Professor of Political Science at Indiana University. His academic interests are in the general area of American politics. His research endeavors have looked at the nature of the linkage between the public preferences of the general public and the policy acts of their representatives. He has published or has articles forthcoming in the *American Political Science Review, American Journal of Political Science, Public Opinion Quarterly, Western Political Quarterly, Legislative Studies Quarterly,* and *Policy Studies Journal.*

STEPHEN H. LINDER received his Ph.D. from the University of Iowa in 1976. He has taught policy and public administration at UCLA and is presently affiliated with the Center for Public Policy Studies at Tulane

University. He is currently interested in modeling decision tradeoffs in environmental policy.

MICHAEL LIPSKY is currently Professor of Political Science, specializing in urban politics and public policy. His recent work has focused on political strategies of relatively powerless groups, and on the politics of public service bureaucracies such as police, schools and welfare offices, that deal directly with citizens. Books he has written or edited include *Protest in City Politics; Law and Order: Police Encounters; Commission Politics: The Processing of Racial Crisis in America;* and *Theoretical Perspectives on Urban Politics.* Professor Lipsky is currently completing work on a book on "street-level bureaucracy."

DAVID NACHMIAS is Associate Professor of Political Science and Urban Affairs at the University of Wisconsin-Milwaukee and Tel Aviv University. He is coauthor of *Research Methods in the Social Sciences* and *Bureaucratic Culture,* and author of *Public Policy Evaluation.* He contributed to scholarly journals including *American Journal of Sociology, Public Administration Review, British Journal of Political Science, British Journal of Sociology,* and *Political Methodology.*

DAVID H. ROSENBLOOM is Associate Professor of Political Science at the University of Vermont. He previously taught at the University of Kansas and Tel Aviv University. During the 1978-1979 academic year he will be a visiting associate professor at Syracuse University. He is author of *Federal Service and the Constitution* and *Federal Equal Employment Opportunity.* Among his coauthored works are *Bureaucratic Culture: Citizens and Administrators in Israel* and *Personnel Management in Government.* He has also written articles appearing in such journals as *Public Administration Review* and the *American Journal of Sociology.*

FLOYD E. STONER is Assistant Professor of Political Science at Marquette University. He received his Ph.D. at the University of Wisconsin-Madison, where he was a Research Assistant with the Institute for Research on Poverty. A study of the passage and implementation of ESEA Title I was supported by a Post-Doctoral Fellowship at Michigan State University. As a consultant to the National Institute of Education, he continued his research on the political role of auditors. Currently he is engaged in a study of the General Accounting Office, supported by a grant from the Marquette Committee on Research.

STEPHEN L. WASBY is Professor of Political Science in the Graduate School of Public Affairs, State University of New York at Albany. For 1978-1979, he is serving as Program Director, Law and Social Science Program, National

Science Foundation, Washington, D.C. He received his Ph.D. from the University of Oregon, and was Russell Sage Post-Doctoral Resident in Law and Social Science, University of Wisconsin at Madison. His books include *The Impact of the United States Supreme Court, Small Town Police and the Supreme Court: Hearing the Word,* and most recently, *The Supreme Court in the Federal Judicial System.* He is a member of the Editorial Boards of *Policy Studies Journal, Justice System Journal,* and *American Politics Quarterly.*

ROBERT K. YIN is a senior research psychologist at The Rand Corporation (Washington, D.C. office). He is also a visiting faculty member at the Department of Urban Studies and Planning, M.I.T. Dr. Yin has published several books on the topic of urban service bureaucracies, including: *Street-Level Governments, Tinkering with the System: Technological Innovations in State and Local Services,* and *Changing Urban Bureaucracies: How New Practices Become Routinized.* He has also published articles in such journals as the *Administrative Science Quarterly, Policy Sciences,* and *Sociological Methods and Research.*